致加西亚的信

全民阅读·经典小丛书

[美]哈伯德◎著

冯慧娟◎编

吉林出版集团股份有限公司

图书在版编目（CIP）数据

致加西亚的信 /（美）哈伯德著；冯慧娟编. —长春：吉林出版集团股份有限公司，2015.6

（全民阅读. 经典小丛书）

ISBN 978-7-5534-7792-3

Ⅰ.①致… Ⅱ.①哈… ②冯… Ⅲ.①职业道德－通俗读物 Ⅳ.①B822.9-49

中国版本图书馆 CIP 数据核字 (2015) 第 128452 号

ZHI JIAXIYA DE XIN

致加西亚的信

作　　者：［美］哈伯德　著　冯慧娟　编
出版策划：孙　昶
选题策划：冯子龙
责任编辑：颜　明　姜婷婷
排　　版：新华智品
出　　版：吉林出版集团股份有限公司
（长春市福祉大路 5788 号，邮政编码：130118）
发　　行：吉林出版集团译文图书经营有限公司
（http://shop34896900.taobao.com）
电　　话：总编办 0431-81629909　营销部 0431-81629880 / 81629881
印　　刷：北京一鑫印务有限责任公司
开　　本：640mm × 940mm 1/16
印　　张：10
字　　数：130 千字
版　　次：2015 年 10 月第 1 版
印　　次：2019 年 6 月第 3 次印刷
书　　号：ISBN 978-7-5534-7792-3
定　　价：32.00 元

印装错误请与承印厂联系　电话：18611383393

前言

《致加西亚的信》是一个通俗易懂的故事，所蕴含的概念直白明了，但它超越了众多大学里教授学生的空洞理论，使“送信”这一行为成为忠诚、责任、敬业和荣誉的象征。世界上许多政府、军队和企业都把这本书作为培养职员、士兵和员工的宝典。在美国著名的西点军校，这本书是训练学员自立和主动性的教材；美国前佛罗里达州州长（1999—2007）杰布·布什第一次竞选州长成功后，就将此书送给了自己的副手——副州长弗兰克·布洛根；众多跨国公司要求员工人手一册的情况，更是数不胜数。

这本书对个人、企业、国家，甚至整个人类文明的发展都产生了重要影响。从它第一次出版到现在，已有一百多年的历史。它在世界范围内广泛流传，一直深受读者的喜爱，成为全球最畅销的图书之一，各种版本加在一起的销售量高达8亿册。

许多人在看过《致加西亚的信》后都认为，这是从老板角度出发而写的书。这种观点其实失之偏颇，因为从员工忠诚和敬业中获益最大的不是公司和老板，而是员工自己。一个人只有对自己的工作持有高度的责任感和忠诚度，才会被他人信任，才会被委以重任；而如果一个人懒惰、抱怨，即使自己创业做老板，也不会取得成功。

好书能提高人们的思想境界，希望读罢此书，读者朋友们能够领悟罗文的职业精神，把“忠诚、责任、敬业、荣誉”的精神贯彻到工作中，让自己的发展有一个质的飞跃。

目录

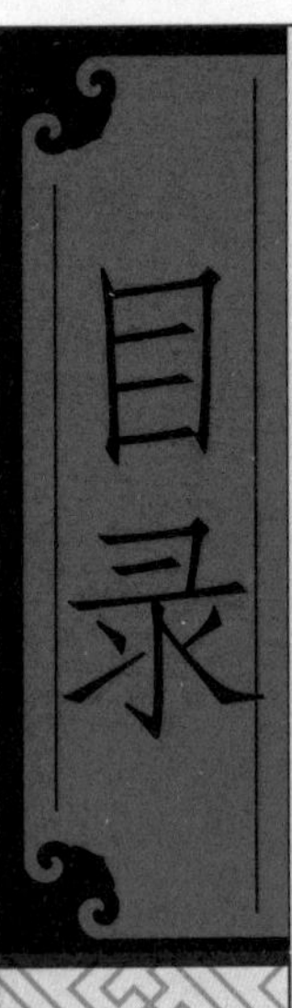

第一部分

1913年版作者序言

Forword For The Version In 1913

我是在一天的晚饭后完成《致加西亚的信》的，仅用了一个小时。那天正好是华盛顿的诞辰，1899年2月22日。当时，我们正热火朝天地准备出版三月份的《菲士利人》，想告诉那些没有责任感的市民要从昏睡中清醒过来，打起精神，振作起来。

《致加西亚的信》的创作灵感来源于喝茶时大家的聊天内容。当时，我的儿子博尔特认为古巴战争中真正的英雄不是加西亚，而是罗文，是他孤身一人完成了将信送给加西亚的任务。

儿子的言论如同一朵火花闪现在我的脑海中！没错，他说得没错，真正的英雄是那些尽职尽责完成了自己工作的人，是把信送给加西亚的人。我来到书桌前，一气呵成地写出了一篇文章，也就是后来的《致加西亚的信》，当时它并没有标题。我将文章刊登在《菲士利人》杂志上。杂志很快售罄，让人吃惊的是，要求加印的订单一波接一波，一打，五十份，一百份……当得知美国新闻公司订购了1 000份的时候，我问助手到底是哪篇文章引起了人们的兴趣，他说：“是一篇关于加西亚的文章。”

第二天，纽约中心铁路局的乔治·丹尼尔发来电报，电报的内容是："把关于罗文的那篇文章单独印刷成小册子，并在册子的封底印上帝国州际快客的广告，现订购10万份，请报价，并告知可以邮寄的日期。"我报了价，因为当时我们的工厂规模很小，印刷10万份小册子对我们而言是一项浩大的工程。所以，我告诉他们这些册子在两年内可以印刷完成。

结果，我答应丹尼尔按照他自己的方式来重新印刷那篇文章。他

竟然卖出去了50万册，其中的两三成都是丹尼尔先生直接发出去的。此外，二百多家杂志和报纸也转载了这篇文章。迄今为止，它已经被翻译成多国文字，传播到世界各个角落。

在丹尼尔先生销售《致加西亚的信》的时候，俄罗斯铁道大臣西拉科夫亲王恰好也在美国。当时，他接受纽约中央铁路公司邀请，在丹尼尔先生的陪同正参观纽约市。所以，他也看到了这本小册子，并且对它产生了极大的兴趣。其中最重要的原因也许是这本小册子非常畅销。总之，西拉科夫将它带回到俄罗斯，并让人把它翻译成俄文，然后给俄罗斯铁路工人每人发一册。

其他国家也纷纷引进并翻译了这篇文章，这些国家包括：德国、法国、西班牙、土耳其、印度、中国。日俄战争爆发后，每位上前线的俄罗斯士兵人手一份《致加西亚的信》。在被俘俄罗斯士兵的身上，日本人发现了这些小册子，他们认为这一定是十分有价值的东西，于是这篇文章又有了日文版。日本天皇还颁布命令，每一位日本政府官员、士兵以及百姓都要有一册《致加西亚的信》。迄今为止，《致加西亚的信》的印数高达4 000千万册。

可以说，在一个作家的有生之年，还没有哪部文学作品拥有过如此巨大的发行量，这幸亏有一系列偶然发生的事件。

阿尔伯特·哈博德

1913年12月1日

This literary trifle, *A Message To Garcia*, was written one evening after supper, in a single hour. It was on the 22nd of February 1899, Washington's Birthday: we were just going to press with the *March Philistine*.The thing leaped hot from my heart, written after a trying day, when I had been endeavoring to train some rather delinquent villagers to abjure the comatose state and get radioactive.

The immediate suggestion, though, came from a little argument over the teacups, when my boy Bert suggested that Rowan was the real hero of the Cuban War. Rowan had gone alone and done the thing—carried the message to Garcia.

It came to me like a flash! Yes, the boy is right, the hero is the man who does his work—who carries the message to Garcia. I got up from the table, and wrote *A Message To Garcia*. I thought so little of it that we ran it in the Magazine without a heading. The edition went out, and soon orders began to come for extra copies of the *March Philistine*. A dozen, fifty, a hundred, and when the American News Company ordered a thousand, I asked one of my helpers which article it was that stirred up the cosmic dust. "It's the stuff about Garcia," he said.

The next day a telegram came from George H. Daniels, of the New York Central Railroad thus, "Give price on one hundred thousand Rowan article in pamphlet form—Empire State Express advertisement on back—also how soon can ship."I replied giving price, and stated we could supply the pamphlets in two years. Our facilities were small and a hundred thousand booklets looked like an awful undertaking.

The result was that I gave Mr. Daniels permission to reprint the article in

his own way. He issued it in booklet form in editions of half a million. Two or three of these half million lots were sent out by Mr. Daniels, and in addition, the article was reprinted in over two hundred magazines and newspapers. It has been translated into all written languages.

At the time Mr. Daniels was distributing *A Message To Garcia*, Prince Hilakoff, Director of Russian Railways, was in this country. He was the guest of the New York Central, and made a tour of the country under the personal direction of Mr. Daniels. The Prince saw the little book and was interested in it, more because Mr. Daniels was putting it out in big numbers, probably, than otherwise. In any event, when he got home he had the matter translated into Russian, and a copy of the booklet given to every railroad employee in Russia.

Other countries then took it up, and from Russia it passed into Germany, France, Spain, Turkey, Hindustan and China. During the war between Russia and Japan, every Russian soldier who went to the front was given a copy of *A Message To Garcia*. The Japanese, finding the booklets in the possessions of the Russian prisoners, concluded it must be a good thing, and accordingly translated it into Japanese.And on an order of the Mikado, a copy was given to every man in the employ of the Japanese Government, soldier or civilian.

Over forty million copies of *A Message To Garcia* have been printed. This is said to be a larger circulation than any other literary venture has ever attained during the lifetime of an author, in all history—thanks to a series of lucky accidents.

Elbert Hubbard

December 1st, 1913

第二部分
致加西亚的信

（阿尔伯特·哈博德）

A Message To Garcia

在所有关于古巴的事件中，有一个人深刻地留在我的脑海中，甩都甩不掉。

美国和西班牙之间的战争爆发后，美国迫切需要和古巴起义军首领加西亚取得联系。加西亚当时隐藏在古巴的大山里，没有人知道他藏身的确切地点，所以无法寄信或发电报给他。可是美国总统必须尽快和他合作，一定要和他取得联系，如何是好呢？

这时有人向总统推荐罗文，说：“他能帮您找到加西亚，也只有他能做到。”

于是，他们找来罗文，交给他一封信，让他务必送到加西亚手中。罗文是如何接过信，把它密封在油纸袋里，放在胸前；如何乘船在古巴海面上漂流了四天；如何登上陆地，消失在丛林中；如何在三个星期中徒步越过这个到处都是危险的国家，将信交给了加西亚。我就不一一叙说这些细节了。我要特别提出的是：麦金莱总统把一封写给加西亚的信交给罗文，罗文接过信后却没有问："他在哪里？"

这种伟大的精神会永远流传世间。我们应该为拥有这种精神的人塑造永不腐朽的雕像，并将雕像放在每一所大学里。年轻人需要的不仅是书本知识，也不仅是认真听取各种教诲，而是需要那些能让他们挺胸抬头的敬业精神。有了这种精神，他们会坚定自己的理想，迅速行动，集中精力，全力以赴地完成任务——"把信送给加西亚"。

虽然加西亚早已离开人世。但是，还有更多"其他的加西亚"。我们经常能看到这种情况：凡是需要众多人手的企业经营者，有时候都会因属下无法或不愿专心去做一件事而大吃一惊。懒散、冷淡、马虎的做事态度，已经成为常态。让属下做事，要么苦口婆心、威逼利诱，要么奇迹出现，上帝派一名助手给他，否则，没有人能把事情办好。

如果你不认同我的结论，那就来做个测验。假设你有六名属下，在你办公时，你找来其中的一名，吩咐道："马上在百科全书中查找科勒乔的资料，并做一份关于他生平的简短备忘录。"这名员工会不会迅速地回答你"是，先生"，然后马上去工作呢？实际上，他不会。他会用怀疑的眼光看着你，然后问你下面这些问题，或者是这些问题中的某些问题，或者是更多的问题。

科勒乔是什么人？

在哪一本百科全书中查找？

百科全书放在哪里了？

我的工作就是做这个吗？

查理为什么不能做这件事呢？

科勒乔去世了吗？

真的很紧急吗？

是否需要我把百科全书找来你自己查呢？

你想知道科勒乔的哪些资料呢？

我敢用十倍于你的赌注和你打赌，当你回答了他提出的所有问题，并告诉了他怎样查找资料以及查找的理由后，这名员工就会走开，然后找另外一名员工帮他的忙。然后，他回来对你说，书里根本没有这个人。当然，我也许会在打赌中输掉，但根据发生的概率，可能性很小。

如果你是个有智慧的人，你就不会不厌其烦地向你的属下解释提出的问题：应该在字母“C”的索引中查找科勒乔的资料，而不是在字母“K”字的索引中。你会平静地说“算了”，然后亲自去查找。像员工这种行为上的被动、道德上的愚蠢、决心的不坚定、不愿意主动做事的作风，很有可能把社会带到三个和尚没水喝的危险境地。人们为自己的利益都不采取积极行动，那么，你又怎能期待他们为公众利益而采取行动呢？你登广告招聘一名速记员，应征者中会有大部分人既不会拼写也不会使用标点，他们甚至不认为这是速记员的必备技能。

你能奢望这样的人把信送给加西亚吗？

在一家规模不小的公司里，一位主管指着一个人对我说：“那是我们的会计。”“我看见了，他有问题吗？”主管无奈地说：“他是一位技能出色的会计师，但是如果我让他去城里办事，他可能很快就完成了任务，也可能在去城里的路上走进4家酒吧，等他到了城里，恐怕早就把任务忘掉了。”你能奢望这样的人把信送给加西亚吗？

最近，我发现很多人都对那些“收入微薄而且永无出头之日的人”和“为求得一份稳定工作四处奔波、无家可归的人”表示出深切的同情。同时，这些人还痛骂雇主。但是，从来没有人提到过，有些雇主尽自己最大的努力都没有能够使那些懒散的职员做些有意义的工作；也没有人提过，有些雇主是如何以自己的耐心去感动那些自己一转身就投机取巧的员工的。

每个企业和商店，对自己所雇佣的员工都有一个持续性的调整过程。雇主会解雇那些对公司没有丝毫贡献的员工，同时招收新的员工。

不管业务有多忙，这种调整工作一直都在进行。在公司不景气，能提供的就业机会很少的情况下，这种调整才会显出较好的成绩——那些不能胜任工作、没有才能的人就会被摒弃在就业大门之外，能干的人会留下来。这也是适者生存的原则。为了保证自己的利益，雇主只能留下最佳员工——因为只有他们才会把信送给加西亚。

我见过一个很聪明的人，但他没有独自创业的能力。对其他人而言，他也没有任何价值。因为他有一个愚蠢的思想，那就是认为雇主在剥削自己，或企图剥削自己。他既没有领导他人的能力，也不甘心让他人领导。如果让他去把信送给加西亚，他极有可能会说："还是你自己去吧！"今晚，他依然到处奔波想找一份工作，寒风毫不留情地钻进他的破外套里。但是知道他性格的人都不愿意给他一份工作，因为这是一个引起不满的导火线。

当然，我们都知道，这种道德有缺陷的人，不会比一个肢体有缺陷的人更值得同情。但是，我们也应该对那些勤奋经营大企业的人洒下同情的泪水，他们不会因为下班的铃声而丢掉手头的工作，他们因为要全力去使那些懒惰、冷漠、被动、没有能力的人不太离谱而日增白发。如果没有这样的勤奋和心思，那些员工就会露宿街头，忍冻挨饿。

我这样说是不是有些夸张呢？可能吧。不过，如果全世界都变成了贫民窟，那么，我要为成功者说几句同情的话——他们在几乎没有成功可能性的情况下全力引导其他人，并最终取得成功。可是，除了食物之外，他们没有获得任何东西。我也曾为了生存替人工作，也曾做过老板让其他人工作，所以知道应该全面地看问题。

贫穷本身并不美好，它也不值得被推介。但也不是所有的老板都贪婪、专横，就好比并不是所有的人都善良一样。我尊重那些不管老板是否在办公室都会勤奋工作的人。我也尊重那些把信交给加西亚的人，他们只会静静地接受命令，不会提出愚蠢的问题，不会出门后把信扔到水沟里，不会不去送信而做其他无关的事情。这些人永远不会被解雇，也永远用不着为了加薪而罢工。

文明，就是焦急地寻找这些人的一个长远过程。这些人不管要求什么东西，最后都会得到。他的才干如此独一无二，不可或缺，任何雇主都不愿失去他。他在每个城市、乡镇、村庄，以及每个办公室、商店、工厂都会受到热烈欢迎。世界需要这样的人才，需要能把信送给加西亚的人。

阿尔伯特·哈博德写于1899年

In all this Cuban business, there is one man stands out on the horizon of my memory like Mars at perihelion.

When war broke out between Spain and the United States, it was very necessary to communicate quickly with the leader of the Insurgents. Garcia was somewhere in the mountain vastness of Cuba—no one knew where. No mail nor telegraph message could reach him. The President must secure his cooperation, and quickly. What to do?

Some one said to the President: "There's a fellow by the name of Rowan who will find Garcia for you, if anybody can. "

Rowan was sent for and given a letter to be delivered to Garcia. How "the fellow by the name of Rowan" took the letter, sealed it up in an oil-skin pouch, strapped it over his heart, in four days landed by night off the coast of Cuba from an open boat, disappeared into the jungle, and in three weeks came out on the other side of the Island, having traversed a hostile country on foot, and delivered his letter to Garcia—are things I have no special desire now to tell in detail. The point that I wish to make is this: McKinley gave Rowan a letter to be delivered to Garcia; Rowan took the letter and did not ask "Where is he?"

By the Eternal! There is a man whose form should be cast in deathless bronze, and the statue placed in every college of the land. It is not book-learning young men need, nor instruction about this and that, but a stiffening of the vertebrae which will cause them to be loyal to a trust, to act promptly, to concentrate their energies: do the thing— "Carry a message to Garcia!"

General Garcia is dead now, but there are other Garcias. No man who has endeavored to carry out an enterprise where many hands were needed, but has been well-nigh appalled at times by the imbecility of the average man—the inability or unwillingness to concentrate on a thing and do it.

Slipshod assistance, foolish inattention, dowdy indifference, and half-hearted work seem the rule; and no man succeeds, unless by hook or crook or threat his forces or bribes other men to assist him; or mayhap, God in His goodness performs a miracle, and sends him an Angel of Light for an assistant.

You, reader, put this matter to a test: You are sitting now in your office—six clerks are within call. Summon any one and make this request: "Please look in the encyclopedia and make a brief memorandum for me concerning the life of Correggio." Will the clerk quietly say "Yes, sir" and go do the task? On your life, he will not. He will look at you out of a fishy eye and ask one or more of the following questions:

Who was he?

Which encyclopedia?

Where is the encyclopedia?

Was I hired for that?

What's the matter with Charlie doing it?

Is he dead?

Is there any hurry?

Shan't I bring you the book and let you look it up yourself?

What do you want to know for?

And I will lay you ten to one that after you have answered the questions, and explained how to find the information and why you want it, the clerk will go off and get one of the other clerks to help him try to find Correggio—and then come back and tell you there is no such man. Of course I may lose my bet, but according to the Law of Average, I will not.

Now, if you are wise, you will not bother to explain to your "assistant" that Correggio is indexed under the C's, not in the K's, but you will smile very sweetly and say "Never mind" and go look it up yourself. And this incapacity for independent action, this moral stupidity, this infirmity of the will, this unwillingness to cheerfully catch hold and lift—these are the things that put pure Socialism so far into the future. If men will not act for themselves, what will they do when the benefit of their effort is for all? A firstmate with knotted club seems necessary;and the dread of getting"the bounce"Saturday night holds many a worker to his place. Advertise for a stenographer, and nine out of ten who apply can neither spell nor punctuate—and do not think it necessary to.

Can such a one write a letter to Garcia?

"You see that bookkeeper," said the foreman to me in a large factory. "Yes, what about him?" "Well, he's a fine accountant, but if I'd send him up town on an errand, he might accomplish the errand all right, and on the other hand, might stop at four saloons on the way, and when he got to Main Street would forget what he had been sent for." Can such a man be entrusted to carry a message to Garcia?

We have recently been hearing much maudlin sympathy expressed for the "downtrodden denizens of the sweatshop" and the "homeless wanderer searching for honest employment", and with it all often go many hard words for the men in power. Nothing is said about the employer who grows old before his time in a vain attempt to get frowsy never-do-wells to do intelligent work; and his long patient striving after "help" that does nothing but loaf when his back is turned.

In every store and factory there is a constant weeding-out process going on. The employer is constantly sending away "help" that have shown their incapacity to further the interests of the business, and others are being taken on. No matter how good times are, this sorting continues: only, if times are hard and work is scarce, the sorting is done finer—but out and forever out the incompetent and unworthy go. It is the survival of the fittest.Self-interest prompts every employer to keep the best—those who can carry a message to Garcia.

I know one man of really brilliant parts who has not the ability to manage a business of his own, and yet who is absolutely worthless to anyone else, because he carries with him constantly the insane suspicion that his employer is oppressing, or intending to oppress, him. He cannot give orders; and he will not receive them. Should a message be:given him to take to Garcia, his answer would probably be: "Take it yourself!" Tonight this man walks the streets looking for work, the wind whistling through his threadbare coat. No one who knows him dare employ him, for he is a regular firebrand of discontent. He is impervious to reason,and the only thing that can impress him is the toe of a thick—soled Number Nine foot.

Of course I know that one so morally deformed is no less to be pitied than a physical cripple; but in our pitying, let us drop a tear, too, for the men who are striving to carry on a great enterprise, whose working hours are not limited by the whistle, and whose hair is fast turning white through the struggle to hold in line dowdy indifference, slipshod imbecility, and the heartless ingratitude which, but for their enterprise, would be both hungry and homeless.

Have I put the matter too strongly? Possibly I have; but when all the world has gone a-slumming I wish to speak a word of sympathy for the man who succeeds—the man who, against great odds, has directed the efforts of others, and having succeeded, finds there's nothing in it: nothing but bare board and clothes. I have carried a dinner pail and worked for day's wages, and I have also been an employer of labor, and I know there is something to be said on both sides.

There is no excellence, perse, in poverty; rags are no recommendation; and all employers are not rapacious and high-handed, any more than all poor men are virtuous. My heart goes out to the man who does his work when the "boss" is away, as well as when he is at home. And the man who, when given a letter for Garcia, quietly takes the missive, without asking any idiotic questions, and with no lurking intention of chucking it into the nearest sewer, or of doing aught else but deliver it, never gets "laid off " nor has to go on a strike for higher wages.

Civilization is one long anxious search for just such individuals. Anything such a man asks shall be granted; his kind is so rare that no employer can afford to let him go. He is wanted in every city, town and village; in every office, shop, store and factory. The world cries out for such: he is needed and needed badly, the man who can "Carry a Message to Garcia."

Elbert Hubbard

1899

第三部分
我是怎样把信送给加西亚的?

（安德鲁·萨默斯·罗文）

How I Carried The Message To Garcia?

“在哪里，”麦金莱总统急切地询问军事情报局局长阿瑟·瓦格纳上校，“在哪里能找到把信交给加西亚的人？”

“在华盛顿就可以找到，他是陆军中尉罗文，他能将信送给加西亚！”上校迅速地回答。

“派他去！”总统下达了命令。

当时，美国和西班牙之间的战争一触即发，麦金莱总统迫切希望得

到敌方的情报。他深知一点，要想战胜西班牙，美国军队必须和古巴的起义军联合起来。他需要掌握西班牙军队在岛上的防御体系：包括士兵的生活环境、士气，以及军官——尤其是高级军官的性格；岛上全年的道路状况；军队的医疗卫生水平和装备情况。他还想知道当美国军队在集结期间，古巴起义军需要什么才能压制住敌军，以及其他众多重要的情报。“派他去！”总统的命令就三个字，犹如上校的回答一样，干脆而迅速。

大约一小时后，正是中午，瓦格纳上校通知我一点钟到陆海军俱乐部，和他一起吃午餐。顺便提一点，这位上校爱开玩笑的性格远近皆知。吃饭的时候，他问我：“下一班前往牙买加的船何时出发？”想到他可能是想开玩笑，为了避免被他调侃，我沉默了一会儿才抬起头告诉他：“明天下午有船，从纽约起航，是一艘名叫安迪伦达克的英国船。”

“你能乘上这艘船吗？”上校很急切地问。

尽管我认为上校只是在开玩笑而已，但还是明确地回答说：“能！”“那就准备坐这艘船出发吧！”上校发出这样的命令。

“罗文，”上校接着说道，“总统已经下达命令，让你去联络，或者说将信送给加西亚将军。他在古巴东部的某个地方，具体在哪里谁也不知道。你的任务就是安全、迅速地把信交给他，并把那里真实而具有实用性的军事情报带回来。总统把他想知道的问题都写在了你带的这封信上。另外，你不能携带任何有可能暴露你身份的东西。历史上曾发生过许多这样的悲剧，我们没有理由再次冒险。大陆军的内森·赫尔，美墨战争中的里奇中尉都在身上带有情报的情况下被捕，而且都牺牲了，作战计划也被敌人得知了。你绝不能失败，一定要确保万无一失。”

此时，我才最终明白，瓦格纳上校并没有和我开玩笑。他继续说：“当你到达牙买加后，会有人找到识别你身份的办法，那里有古巴游击队。剩下的事情就全靠你自己了，不过你不会得到比现在更多的信息。”确实是这样，他现在所说的这些话也只是轮廓而已。“你马上去做准备，军需官哈姆菲里斯会把送你到金斯敦。一旦我国和西班牙交战，那么，我们的战略指示都将以你发来的电报为依据，否则我们对敌方一无所知。这项任务意义重大，你自己全权负责，一定把信送给加西亚将军。火车在午夜时分离开，祝你好运，再见！”上校交代完毕，我们握手道别。

瓦格纳上校送我离开的时候再三叮嘱：“一定要把信送给加西亚将军！”我一边忙着准备出发，一边考虑这项任务。我明白任务的艰巨性和复杂性。现在美西之间还没有开战，也许我出发的时候战争也没有爆发，甚至我到达牙买加时都不会爆发。但只要走错一步都会为自己带来可能一生都无法解释清楚的麻烦。如果双方开战了，任务反倒轻松了一些，尽管危险性并没有减少。在这种情况下，一个人的名誉和生命都处在极度危险中，寻求指示变得理所当然。军人的生命被国家掌控着，但荣誉却属于他自己。生命可以丧失，但荣誉不能落到任何会破坏它的人手中，也不能被忽视或遭到轻视。这一次，我无法按照任何人的指令行事，一切都只能靠自己。我必须把信送给加西亚将军，并从他那里获得国家需要的情报，这是我一定要做到的事情。我不知道瓦格纳上校是否把我们谈话的内容记在了办公室的档案中。与将要开始的任务相比，这已经不是重要的事了。

必须乘坐的那趟火车在午夜零点零一分开车，让我想起了一个流传很久的迷信说法，星期五不是出门的好日子。虽然火车离开华盛顿之时，已经是星期六了，但是我出门的时间还是星期五。我想，这就是命运刻意而为的吧。不过，当我开始思考其他事情的时候，就忘了这个说法，一直将它抛在脑后，现在这个迷信已经没有什么意义了，因为我的任务已经圆满地完成了。

我赶上了那艘开往牙买加的轮船“安迪伦达克”号，幸运的是它准时起航，而且一路上平安无事。在船上，我尽量避免与其他人接触。不过，后来结识了一名电机工程师，我俩成了旅途上的伙伴。从他的嘴里我知道了自己的形象，总是独来独往，从来也不说自己的情况，所以几个爱开玩笑的家伙就给我起了一个外号——“冷漠的人”。

当“安迪伦达克”号驶入古巴海域的时候，我意识到，危险即将到来。我身上有一封美国国务院写给牙买加官方的信件，用来证明我的身份。但是，这个文件也很可能让我在敌人面前暴露身份。如果战争在“安迪伦达克”号驶入古巴海域之前就开始了，那么，西班牙人就可以根据国际法到船上搜查。对他们而言，我既是非法入境者，也是非法送信人，所以他们会把我当作战犯逮捕起来。而且，这艘英国轮船也很有可能被击沉，尽管它悬挂着中立国的国旗，从一个和平的港口驶往一个中立国的港口（牙买加当时是中立国），并不知道战争随时都会爆发。

想到这里，我意识到情况很严重，于是把那个文件放在特等舱的救生衣里藏起来。直到“安迪伦达克”号远离了这片海域，我才放下一直悬着的心。第二天早上九点，船终于到达目的地，我踏上了牙买加的国土，并设法和古巴的游击队队长莱恩先生取得联系。我，莱恩，以及他的助手一起商讨如何将信迅速地送给加西亚将军。我在四月九日离开华盛顿，四月二十日，我看到美国发来的电报，说西班牙已经同意在二十三日之前将古巴交还给古巴人民，并撤走岛上的所有武装力量以及海域上的海军。我用密码向政府发出我得到的消息，四月二十三日我收到密电：“尽快见到加西亚将军！”

接到密电几分钟后，我来到游击队指挥中心，他们正在等我。当时，那里还有一些我之前从未见过的流亡在外的古巴人。正当我们讨论如何送信时，一辆马车从外面驶进来。

“时间到了！”有人用西班牙语喊着。

还没来得及有任何反应，我已经被带上了那辆马车上。

就这样，一个军人服役以来最为惊险的一次经历就这样开始了。驾驶马车的人大概是世界上最不爱说话的马车夫。他不但不与我主动说话，就是我说话他也像没听见一样，丝毫不做回应。从我被带上车那一刻起，他就驾驶着马车在迷宫般的金斯敦大街上飞驰，速度丝毫不减。很快我们就来到了郊区，将整个城市远远地抛在后面。我敲了敲车，甚至还踢了一下，可是，马车夫依然没有反应。

他似乎知道我的任务，而他的职责就是用最快的速度让我走完第一段路程。好几次，我都努力让他听听我的话，可惜都以失败告终。最后，我无奈地坐回座位，顺其自然吧。

大约又前行了四英里，我们穿过一片浓密的热带森林，然后沿着一条平坦的西班牙小镇公路向前行驶，最后停在一片丛林的边上。马车的门从外面被打开了，我看见了一张并不熟悉的脸。紧接着，我被转入另一辆车，看来它已等候多时。真是奇怪啊！好像每件事情都做了精心的安排，没有多余的话，而且一秒钟都不用浪费。

就这样，我又马上开始了下一段旅程。第二个马车夫与第一个一样都默不作声。尽管我尽一切努力想和他说几句话，但都没有得到他的回应。他只管驾驶着马车飞速前进。很快，我们穿越了西班牙小镇，来到克伯利河谷，然后又进入岛屿中央，那里有一条直通圣安斯湾加勒比海水域的路。

后来，我催促马车夫和我说话，可他依然沉默不语。他好像听不懂我说的话，也不明白我手势的意思，他唯一做的就是驾车前进。随着所走的路地势逐渐增高，我的呼吸畅快了许多。太阳落山时，我们在一个火车站旁边停了下来。我发现，山坡上有黑乎乎的东西朝我翻滚过来，那是什么？难道西班牙当局已经知道我的到来，于是安排牙买加军官跟踪并追赶我?

看到这个黑色幻影的时候，我的心中一阵不安。不过，我随后松了一口气，因为我发现那是一位步履蹒跚的老黑人。他向我递上了可口的炸鸡和两瓶巴斯啤酒。此外，他还不停地对我说话，可惜是一口当地的方言，我只能听懂几个词。不过，我知道他是在表达他的感激之情，感谢我帮助古巴人民赢取自由，他给我食物只是想表示自己的心意。

此时，马车夫犹如一个局外人，他对食物没有兴趣，对我和老黑人之间的谈话也不感兴趣。时间不长，我们换过两匹马后又要出发了。车夫用鞭子抽打着马，我赶紧与那位年老的黑人告别："再见，老人家！"不一会儿，我们已经把他远远地抛在身后，消失在了苍茫的夜色中。虽然我明白自己所肩负的重担，刻不容缓地赶路，但是，当见到夜晚中的热带雨林后，我还是被它吸引了。在白天，热带雨林是一个翠绿的植物世界；到了晚上，这里就成了昆虫的世界，有各种各样的虫子飞来飞去。萤火虫已经点亮了自己的磷光灯，忽上忽下地飞舞着，有种奇异的美。当马车穿越时，萤火虫的光照亮了整片雨林，我仿佛进入了一个仙境。

可是，想到自己必须要完成的任务，也只能将这美景抛在脑后。马车继续飞奔而行，不过马的体能消耗很大，逐渐体力不支了。就在这时，一声刺耳的哨声从灌木丛中传出来，马车也停止了行驶。一群全副武装的人好像从地底下钻出来的一样，一下子就把我们包围了。

在英国管辖的地方遭到西班牙士兵的拦截并不可怕。不过，这次毫无预兆的停顿让我感到紧张。如果牙买加当局得到消息，知道我违反了这个小岛的中立原则，他们就会阻止我前行。这样，我的任务就会因为牙买加政府的干预而失败。这些人要是英国士兵该多好啊！幸运的是，我的忧虑很快就消失了。因为一番谈话之后，我们被放行了，又可以上路了。

大概一小时后，马车停在一座房子前，房间里闪着昏黄的灯光，一顿丰盛的晚餐正等着我们。这是游击队为我们准备的，他们认为人就应该肆无忌惮地吃美味的食物，并递给我一杯牙买加朗姆酒。虽然我们已经在路上飞驰了九个小时，走了七十多英里路，还更换了两班人马，但是，我并没有感到疲惫，只觉得这杯朗姆酒芳香无比，让人愉悦！接下来，我们分别进行自我介绍。这时，从隔壁屋子里走出来一个人，他身体强壮，看上去性格坚毅。他留着长须，有一只手断了一根手指。这是一个在任何时候，包括紧急时刻都可以让人信任的人。他的眼神忠诚而热情，显示出高贵的品质。他是一位西班牙人，曾经去过古巴。在圣地亚哥时，他因为反抗西班牙旧制度而被砍了一根手指，后来被流放到这里。他叫格瓦希奥·萨比奥，是我这次行动的向导，护送我直到我把信交到加西亚将军手中。另外，他们还雇请当地人将我送出牙买加，这些人再向前走7英里就算完成任务了。不过有一个人例外，那就是我的“助手”。

在这里休息了大概一个小时，我们又出发了。走了半个小时后，又有口哨声把我们拦住了。下了车，我们进入一片甘蔗园，在里面悄无声息地穿行，走了大概一英里路，来到了一个小果园，里面全是可可树。小果园离海湾非常近，而距离海湾五十码的地方就有一艘轻轻晃动的渔船。

这时，船里闪出一丝亮光，我猜这是一个报时的信号。因为我们的行动一直悄无声息，不会有其他人发觉的。格瓦希奥显然对船上人的警觉性很满意，于是作出了回应。接着，我对游击队派来的人表示感谢。然后爬上一名船员的背，他趟水过来把我接上船。至此，我有惊无险地完成了给加西亚将军送信的第一段路程。

到了船上，我发现了很多用来压舱的大石块；长方形的东西是一捆一捆的货物，它们都不足以影响船的行驶。但它们占了很大的空间，再加上两名船员，使船上的空间显得比较狭窄，在里面待着并不舒服。格瓦希奥是船长，我告诉他，我非常希望能快速地走完剩余的三英里路。因为他们提供的帮助热情而周到，我深感过意不去。可惜的是，由于海峡过于狭小，无法提供足够大的风力让我们的船前行，我们必须绕过海峡。船很快绕过了海峡，还赶上了微风。险象环生的第二段行程就这样开始了。

毫不隐瞒地讲，出发后我有过十分紧张的时候。如果在离牙买加海岸只有三英里的地方被敌人逮住，那么，我的名誉将荡然无存；如果在离古巴海岸只有三英里的地方被敌人逮住，那么，我的生命将岌岌可危。我的朋友除了这些船员，就是加勒比海了。

这片水域向北一百英里是古巴海岸，那里有西班牙轻型驱逐舰在巡逻。舰上有小口径的枢轴炮和机枪，船员们使用的都是毛瑟枪。与他们的武器装备相比，我们的太落后了，不过，这是我后来才知道的。我们船上的武器都是拼凑而来的，乱七八糟地不成体系，在任何地方都能捡到。如何真的和西班牙人遭遇，我可能会命丧于此。但我必须成功，一定要找到加西亚将军，把信交到他手上！

按照行动计划，在天黑以前，我们必须待在距离古巴海岸三英里的地方，然后以最快的速度航行到某个珊瑚岛后面，在那里等到天亮。即使被发现了，因为我们身上并没有证明文件，所以他们不会找到任何证据。即使他们真掌握了什么证据，我们可以把船凿沉。装满石块的船沉下去很快，他们没有办法找到尸体。计划实施得一直很顺利，并没有可怕的情况出现。在空气清爽宜人的清晨，我正准备眯一会儿，以缓解一下紧张的情绪和疲劳的身体。突然，格瓦希奥大叫一声，以致我们全都站了起来，一艘西班牙驱逐舰正朝我们驶来。

驱逐舰上的人用西班牙语命令我们立刻停下。我们只好把帆降下来，然后躲到船舱里面，除了格瓦希奥。他以一副慵懒的姿态靠在舵柄上，使船头和牙买加海岸保持平行。“他们也许会觉得我是一个从牙买加来到这里的‘孤独渔夫’，会让我们走的。”他冷静地安慰我们，并想着对策。事实真如他所言，当驱逐舰来到我们附近时，那位冒失的年轻舰长用西班牙语喊道：“钓到鱼了吗？”

格瓦希奥用西班牙语回答说：“还没有呢，这些讨厌的鱼今天早上就是不上钩！”

假如那位海军少尉（也许是其他军衔）稍微机灵一点儿的话，就一定会发现船上的破绽，我们就会成为他们的“大鱼”。而我也不可能写这个故事了。当驱逐舰远离我们后，格瓦希奥让船员重新吊起船帆，并对我说：“如果您累了想睡觉，现在就可以放心地去睡了，危险已经过去了。”

接下来的六个小时里我睡了个安稳觉，没有受到任何干扰。实际上，即使有，我也不知道。要不是热带明亮的阳光把我从床垫上晃醒，可能我还会睡下去。那些古巴人用他们颇感自豪的英语向我打招呼："睡得好不好，罗文先生？"整个牙买加都被耀眼的阳光晒红了，仿佛是翡翠中的一颗宝石。天空万里无云，犹如绿宝石一般。岛的南面是一片热带雨林，就像一幅美丽的风景画，北面却比较荒凉。古巴的上空笼罩着一片乌云，我们焦虑地看着它，但它并没有散去的迹象。不过，开始有风了，而短时间内，风力已经越来越大，正适宜我们的船航行。格瓦希奥嘴里叼着雪茄，和我们开着玩笑，气氛很轻松。

大概在下午四点的时候，乌云散去了，金色的阳光撒在西拉梅斯特拉山上（此山是岛屿的主山脉），使山显得既庄严又美丽。我们仿佛推开了艺术殿堂的大门，一切不可思议的美景都呈现在眼前。这里花团锦簇，山海相依，海天一色，所有的景物交相辉映，构成了一幅精妙绝伦的画卷，世界上再也找不到比这更漂亮的景色了。在海拔8000英尺的顶峰，竟然有葱茏而雄伟的绿色城堡长达数百英里！

不过，我的惊叹持续时间并不长。格瓦希奥下令收帆减速，我不解其意。他解释道："我们现在的位置比我原先想象的离战区近多了，我们要充分利用海上的优势，避开敌人，保存实力。再往前走，就有被敌人发现的危险，我们没有必要冒这个险。"

我们急忙检查手中的武器。我只有一把斯密斯·维森左轮手枪，于是他们又发给我一支样式恐怖的来福枪。也许曾经有人用过它，但我十分怀疑它现在能否为我们效力。船上人手一把这样的武器，包括我的助手在内。人们都尽职尽责地护卫着桅杆，还可以随手拿起武器。这次任务中最严峻的时刻到来了。之前发生的事都比较顺利，相对而言还算安全。可是现在，危险正悄悄而来。这是一种死亡的危险，只要被捉住，就意味着失去了生存的机会，就意味着无法将信送到加西亚将军的手中。

这时，我们距离海岸大概有25英里，可看上去是那么近。直到午夜，我们才继续行动。此时，海水已经很浅了，人们开始用桨划船。突然一个巨浪卷过来，我们没有花费多大力气，船就在巨浪的推动下，躲进了一个隐蔽而安静的小海湾。在黑暗的掩护下，我们把船停在了离海岸只有50码的地方。我提议马上上岸，遭到了格瓦希奥的反对。他说："此时岸上和海上都有我们的敌人，最好的办法是原地不动。如果有驱逐舰想知道我们的消息，他们就会登上我们经过的那个珊瑚礁，那时我们上岸也不晚。只要穿过昏暗的葡萄架，我们就能无所忌惮地行走了。"

笼罩在天边的热浪已经逐渐退去，成片的葡萄树、红树林、灌木丛和棘树开始显现出轮廓，它们几乎长到了海边。虽然这一切看得并不是很清楚，但具有一种朦胧的美。终于，明亮的太阳从图尔基诺峰上升起来了，把光辉洒在古巴的最高点上（图尔基诺峰是古巴的最高山峰）。刹那间，万象更新，雾霭消失了，笼罩在灌木丛中的黑影不见了，拍打着岸边的灰暗海水也魔术般地变成了绿色。这是一次显著的胜利：光明战胜了黑暗。

船员们都忙着把东西搬到岸上。我站在那里，沉默无语，脑海中出现一位诗人的诗句："黑暗的蜡烛已经燃尽，欢快的白天踮起脚尖站在雾霭茫茫的山顶上。"想必这位诗人也看见过与此时相似的情景。格瓦希奥见我一动不动，认为我可能是累了，于是小声地对我说："先生，那是图尔基诺峰。"

在这样一个美妙的早晨，我伫立在岸边，不禁心潮起伏，仿佛在我的面前有一艘巨大的战舰，上面刻着我最崇拜的人——美洲的发现者哥伦布的名字；一种庄严的使命感油然而生。

很快，我的想象就结束了。货卸完后，我们就上了岸，把船拖到一个狭小的河口里，然后将其反扣，并把它藏在丛林里。这时，一群破衣烂衫的古巴人聚到了我们上岸的地方。他们从何处而来，如何知道我们是自己人？对我而言，这些问题一直是个谜。可以确定的是，他们已经对上暗号了。这些人虽然将自己打扮成搬运工人，不过，我还是看出他们身上有当兵的印记。比如，有些人身上的疤痕是因为被毛瑟枪的子弹击中而留下的。

我们上岸的地方是各条道路的交会点，从此地可以通过各个丛林。向西走约一英里，就可以看到丛林当中有烟柱袅袅地升起。我听说这些烟柱是从古巴难民熬盐用的大锅中冒出来的。那些难民从地狱般的集中营逃出来后就躲进了深山里。

我的第二段行程也结束了。

前面的行程可以说一直是有惊无险，从此刻开始，真正的危险来临了。西班牙军队正在进行疯狂的大屠杀，针对的就是古巴人。这群刽子手由韦勒带头，极其凶残，可以说毫无人性。他们不放过任何一个人，

不管是有武器装备的士兵，还是从集中营逃出来的手无寸铁的难民，统统都杀掉。我明白，把信给加西亚将军的路会更加难走。不过，我顾不上考虑这些，因为必须马上赶路!

此地的地形并不复杂，通往北部的地方有一片大约蔓延一英里的平坦土地，但被丛林覆盖着。男人们要做的就是开路，古巴的路网如同迷宫，也只有这些古巴人才能在迷宫中开出一条路来。烈日炎炎，我如同被炭火烧烤着。看到同行的伙伴没有人穿着多余的衣服，这让我心生羡慕之情。

我们继续前行，海和山遮住了我们的视线，浓密的叶子、曲折的小路、灼热的阳光，使我们每前进一步都要付出巨大代价。这里到处是青翠的灌木丛，但离开岸边来到山脚下，就看不到这样的景色了。我们很快就来到了一个空旷的地方，竟意外地发现了几棵椰子树，树上长满了椰子。椰子的汁非常新鲜清凉，对于干得快要冒烟的喉咙来说，它就是灵丹妙药。不过，此地不能久留。因为前面还有好几英里的路程。在夜幕降临之前，我们需要翻过几个非常陡峭的山坡，到达另一个比较隐蔽的空地。值得高兴的是，我们进入了一片真正的热带雨林。在这里，前进的路比较好走，人的呼吸也比较顺畅。尽管微风似有似无，但仍然让人有一种神志清爽的感觉。

走过这片雨林，我们来到了古巴伯蒂罗到圣地亚哥的“皇家公路”。就在我们逐渐靠近公路的时候，我发现同伴们一个个都消失在了丛林中，只剩下我和格瓦希奥两人了。我转过身去想问他怎么回事，却看到他把一根手指放在嘴边，示意我不要说话，尽快把枪准备好。接着，他也消失在丛林中了。

很快，我就明白了他们这样做的原因。因为耳边传来了响亮的马蹄声，以及西班牙骑兵所携带的马刀发出的“卡塔卡塔”声音，中间还夹杂着命令的声音。如果不是同伴们高度的警惕性，也许我们早已踏上公路，不幸与敌人短兵相接！我迅速地竖起来福枪，并将我的斯密斯维森左轮手枪也准备好，焦急地等待下面发生的事。我很希望能听到枪声，可是没有听到。我的同伴们又一个个地回来了，格瓦希奥是最后回来的。

“分散开的目的是麻痹敌人，不被他们发现。我们分头行动，彼此距离较远，假如开火，敌人一定会认为这是我们设下的圈套。这本来是一次打击敌人的绝好机会。”格瓦希奥不无遗憾地说，“但是，使命第一！”他笑了笑，然后说，“游戏第二！”

在起义军经常出没的地方，人们有个习惯：他们点起火，在灰里埋上马铃薯。这样，有饥饿的队伍路过就可以拿出来吃。那天下午，我们恰好发现了一个这样的火堆。我们都吃到了喷香的烤马铃薯。然后埋上火堆，继续前行。

在吃马铃薯之时，革命时期的英雄事迹涌现在我的脑海中，马里恩和他的士兵在打仗时也吃烤马铃薯。于是，我的心中出现了这样一种想法：既然马里恩和他的士兵取得了最后的胜利，那么这些古巴人的斗争也能以胜利告终。因为在他们的内心，一种争取民族解放的精神鼓舞着他们，就像这种精神曾经鼓舞了我们国家的爱国志士一样。想到自己的任务就是把信送到他们的将军手中，尽最大的努力让我们国家的军队帮助他们争取民族自由，我顿时感到无上的光荣。

一天的行程结束了，我发现了一些打扮怪异的人。

“那些是什么人？”我发出疑问。

“他们是从西班牙军队里逃出来的士兵，先生。”格瓦希奥回答道，“他们来自曼查尼罗，据他们讲，因为饥饿和残酷虐待才使他们在军队里无法继续待下去。”

逃兵在某些情况下是有用的。但此时，在这旷野中，我对他们持怀疑态度。没有人敢保证他们当中没有奸细，或许有人会溜出去向西班牙军队告密，说一个美国人正试图穿越古巴，目的地很有可能是加西亚将军的营地。敌人不是正想方设法阻挠我完成任务吗？所以，我对格瓦希奥说：“务必要详细盘问这些人，并看住他们，别让他们离开我们的视线！”

“好的，先生。”他一口答应。

事实证明，我的命令是正确的，幸亏我下达了这样的命令。的确有人想逃出去向西班牙军队报告。虽然在怀疑逃兵时，我并没有任何证据，怀疑显然对他们不公平。但是，有两个人最终被证实是间谍。而且，有一个人差一点儿将我杀死。那两个人本来想在黑夜的掩护下离开营地，越过丛林，然后找到西班牙军队，说有一个美国人正试图穿越古巴。

半夜时分，我突然被哨兵的问话声惊醒，然后听到一声枪响。几乎就在同时，一个人影出现在我面前。我本能地一跃而起，却看见对面又出现了一个人影。我还不知道发生了什么，只见第一个人被大刀砍倒了。他的伤口从右肩一直砍到肺部。这个人在临死前供认，他和另一个同伴商量好了，如果同伴没有顺利逃出营地，他就负责将我杀死，目的就是阻止我完成任务。因为他的同伴被哨兵打死了，所以他才来刺杀我。

第二天很晚的时候，我们才将马匹和马鞍准备完毕。在准备的过程中，我们没有办法前进。因为行程被耽搁，我有些烦躁，但也无可奈何。马鞍似乎比马还难寻找，搅得我心烦意乱，于是询问格瓦希奥，说："为什么非要准备马鞍呢？直接骑马就可以了！"

"加西亚将军此时在围攻古巴中部的巴亚莫，先生。"他回答道，"我们所需要走的路程非常遥远。"这就是只有马匹还不够的原因，我们必须找到马鞍和马饰。

有位同伴根据我的马，分给我一副马鞍。在行程过程中，我很快就开始敬佩向导的智慧，这种敬意随着所走路程的增加而增长。假如当时不寻找马鞍，我恐怕要经历一次非常可怕的酷刑，幸好有马鞍为我省去了这种痛苦。我要夸赞这匹马，它套上马鞍和马饰后，比美国平原上任何一匹精心饲养的马匹都要好。

离开了营地，我们沿着山路继续前行。这里的山路弯曲如迷宫，假如一个不熟悉此处地理环境的人，一定会因为迷路而陷入困境。让人称奇的是，我们的向导对这些迂回的路一清二楚，走在其中就像走在大路上一样。

我们离开了一个分水岭，开始从东坡往下走，这时突然遇到了一群服饰鲜艳的小孩和一位白发披肩的老人，他们向我们问好。于是队伍停止前进，格瓦希奥和老人交谈了几句。顿时，“万岁”的呼声响彻天空，他们是在为美国和古巴而欢呼，也为“美国特使的到来”而欢呼。此时情景真让人感动。我一直不清楚他们通过什么途径知道我的到来，但消息快速在丛林中传播。显然，我的出现让这位老人和这些孩子非常高兴。

在亚拉，有一条河沿山脚流经这里。我们在此地宿营。很快，我发现我们再次进入一个危险地带。这里建造了很多“防护墙”，也叫“战壕”，可以阻止西班牙军队从曼查尼罗攻进来，保护峡谷。在古巴，亚拉是一个有着光辉历史的名字。这里是1868至1878年“十年战争”的发源地，第一声对“自由”的呼唤就是从这里发出的。同伴们把我的吊床挂在一堵防护墙的后面。需要说明的是，防护墙并不是真正意义上的战壕，而是不到一人高的石墙。同伴们还找来了一个士兵，让他整夜守护这堵墙。格瓦希奥这样做是为了保护我的安全，以便我能够完成任务。

第二天一早，我们开始攀登西拉梅斯特拉山的北坡，这也是河流的东岸。我们沿着被风化了的山脊赶路。低洼处也许就潜伏着危险，敌人很可能在这里设下埋伏。我们也许会和西班牙的流动部队展开枪战，甚至有被他们逼入困境的危险。

我们沿着河岸，走在坎坷不平的小路上。我一生中曾不止一次见到过虐待动物的野蛮行径，但都没有这次残忍。为了让马走到峡谷的底部，然后再爬上去，我们狠心地抽打着它们。这也是没有办法的事情，因为把信送给加西亚将军是必须要完成的任务。这是战争期间，无数人的自由和生命处在危险之中，几匹马受点折磨又算什么呢？我虽然为它们的遭遇感到难过，但是现在却没有时间哀叹。

万分庆幸的是，最困难的一段行程总算告一段落。在基巴罗森林边缘、一个被玉米地包围的小木屋前，我们停了下来。小木屋的椽子上悬挂着刚砍下的新鲜牛肉，厨师在屋子外面的空地上做饭，欢迎远道而来的美国使者。早有人提前通报了我的到来。在这顿丰盛的大餐中，不仅有鲜嫩的牛肉，还有可口的木薯面包。

刚吃完饭，就听到远处一阵骚乱，森林的边缘传来了说话的声音和马蹄声，原来是卡斯楚上校到了，他是瑞奥斯将军派来的。卡斯楚上校代表他的上司欢迎我的到来，并告诉我，将军会在明天一早到达这里。通知完毕，他敏捷地跳上马，如运动员一样轻快，然后轻扬马鞭，风一般地离开了，就像来时一样快。

卡斯楚上校的到来使我确信一点，那就是我的向导的确经验丰富。

第二天，瑞奥斯将军来了，为我带来了一顶标有“古巴制造”的巴拿马帽子。

瑞奥斯将军被人们称为“海岸将军”，他皮肤黝黑，兼具印第安人和西班牙人的血统。他身形强壮，步履矫健，作战勇猛。在他掌控的地区，西班牙军队的多次进攻都以失败而告终。他的情报来源和直觉一直为人称道。转移并保护隐藏的家属是一件十分困难的事情，但是，他却不可思议地完成了。而且在完成过程中，掌握了敌人的活动情况。西班牙人要采取行动，深入森林，除掉他们。但是，西班牙人始终捉不到他，他们所做的一切努力都徒劳无功。此外，瑞奥斯将军还擅长游击作战，他的军队有时偷袭西班牙的巡逻队，有时近距离狙击，给敌人以沉重打击。

为了保证我的安全，瑞奥斯将军留下两百名骑兵，让他们护送我直至找到加西亚将军。如果此时，有人能看到我们的队伍，就会惊叹，这是一支庞大的队伍。而且，我必须得承认，这也是一支曾经接受严格训练、有快速行军能力的队伍。很快，我们再次进入森林，在西拉梅斯特拉山上的树木掩护下行走。相对而言，这条路上的坑坑洼洼要少一些。不过，这些路非常狭窄。树枝会挡住我们的去路，甚至划破皮肤，把我们的行李弄下马背。但是，让人吃惊的是，向导似乎丝毫不受树枝的影响，步伐稳健。我一般都是在队列中间，有时真想追上他，走在他的前面，然后认真观察他跋山涉水的英姿。他叫迪奥尼思妥·罗伯兹，皮肤又黑又亮，好像煤炭一样，是古巴军队里的一名中尉。他骑着马灵活地穿梭在森林中，根本不把缠结在一起的蔓藤当回事。他用刀为我们在丛林中开出一条路，蔓藤纷纷倒向他的两边，狭窄的小路宽阔了许多。不仅他的刀术令人称奇，他的力气仿佛永远都用不完似的。

4月30日的晚上，我们来到了瑞奥布伊。它位于富饶的巴亚莫河河畔，离巴亚莫城只有20英里。晚上，我们把吊床拉好，刚要准备入睡的时候，格瓦希奥笑容满面地来了，用兴奋的声音说："他在这里，先生！加西亚将军就在巴亚莫。西班牙军队已经撤到了考托河一侧，他们最后一道警卫在考托内河码头的高架上。"

我迫切地想和加西亚将军联系上，于是提议连夜前行。经过一番讨论，他们认为这样做并不稳妥。

1898年5月1日是一个不寻常的日子。当我在古巴的雨林中熟睡的时候，美国海军上将冒着枪林弹雨进入马尼拉湾（当时为西班牙殖民地），进攻西班牙舰队。就在我给加西亚将军送信的路上，美国军队击沉了西班牙战舰，对菲律宾首都形成巨大的威胁。

第二天凌晨，我们就继续赶路了。走过层层梯田，我们来到了巴亚莫平原。这块曾经富裕的土地，因为战火而被抛弃，如今已是荒凉一片，再也看不到繁荣的痕迹。这荒凉之地就是西班牙军队行凶作恶的最好证据。不管怎样，我们终于进入了平原地带。在此之前，我们骑马走了一百多英里荒无人烟的路程，仿佛人类从来没有在这片大自然最青睐的地方出现过。平原上的草丛高大而茂盛，它们既隐藏了我们这支队伍，也为我们遮挡着热辣的阳光。我一想到目的地就在不远的前方，送信的任务即将完成，所有的艰辛和痛苦都被抛在了脑后。就连精疲力竭的马仿佛也能感受到我们这种渴望而急切的心情。

我们在曼查尼罗至巴亚莫的“皇家公路”上，遇见了众多衣不蔽体但兴高采烈的人，他们正着急地向城里走去。这群人一边赶路，一边叽叽喳喳地交谈，这让我联想到自己在丛林中遇到的鹦鹉。他们曾经被驱逐出家园，现在终于可以回去了。

从此处到城里的路途并不算远。巴亚莫这个城市本来人口多达3万，但现在只有约2 000人，变成了一个小村庄。西班牙人在巴亚莫河的两岸修建了许多碉堡，这些碉堡将这个城市包围起来。我们来到这里，最先看到的东西就是它们，而且，里面的烟火还没熄灭。当古巴人重新回到自己曾经富裕的家园时，西班牙人就将这些碉堡全部付之一炬。

我们整齐而迅速地在岸边列好队伍。格瓦希奥和罗伯兹与士兵讲完话后，我们又继续前行。在河中央，我们停了下来，让马喝足水，自己也养精蓄锐，为到达胡卡洛—莫隆的东岸，也就是古巴军队指挥官的营地而积攒力量（引用当天报纸发布的消息就是：古巴将军说罗文中尉的到来在古巴军队中引起了巨大的轰动。罗文中尉骑着马，在向导的陪同下来到古巴）。

几分钟后，我来了加西亚将军的驻地。

漫长而惊险，随时都可能失败、甚至死亡的行程终于结束了。

而我，成功了！

我来到加西亚将军的指挥部门前，看到一位心思细腻的士兵正把一面古巴国旗挂在了指挥部的门上。在战火纷飞的时期，我代表美国政府，将要见到一位传奇式的人物，这让我感到十分激动。

我们从马上下来，整齐地排成一队。格瓦希奥与将军是旧识，所以他先走到门前，卫兵放他进去了。时间不长，他和加西亚将军一同走出来。将军对我的到来表示了诚挚的欢迎，并邀请我和我的助手去屋里谈话。将军把我介绍给自己的部下。这些军官身着白色的军装，腰间佩带武器。将军解释说，他之所以迎接晚了，因为他在审查能够证明我的身份的文件，这些文件是牙买加的古巴联络处送来的。他需要对这些文件进行基本的审查。

幽默无处不在。联络处送过来的文件中将我称作“密使”，而翻译人员却把这个称呼翻译成了“一个自信的人”。吃过早饭，我们开始正式讨论。我告诉加西亚将军，我的使命完全是军事任务，尽管我离开美国时带的是外交书信。但是，美国总统和作战部最想知道的是古巴东部军事情况的最新消息。（曾有两名军官被派往古巴中部和西部，但他们并没有到达目的地，也没能完成任务。）美国有必要了解西班牙军队占领区的详细状况，包括自然环境、军队数量，指挥官，尤其是高级指挥官的性格特征，西班牙军队的士气，以及整个国家和每个地区的地形、通信状况和道路状况。总之，我们必须知道与美国制定作战计划有关的任何讯息。还有很重要的一点就是，美国军队和古巴军队之间以何种方式合作？是共同作战，还是各自为战？加西亚将军的计划是什么？此外，我对加西亚将军说，我们的政府也非常想了解古巴军队的兵力情况，以及是否需要帮助。最后，我表示自己愿意留下来了解情况，希望将军可以为我安排任务。

加西亚将军听完我的讲述后沉默不语。一会儿，他和所有的军官都退下了，屋子里只剩下他的儿子加西亚上校和我。大概在三点钟的时候，将军回来了，并告诉我他的决定，那就是派三名军官陪我回美国。这三名军官都是土生土长的古巴人，十分了解自己的国家。并且他们都训练有素，知识渊博，能回答美国作战部提出的以上所有问题。即使我在古巴待上一段时间，也不可能做出比他们更完美的报告。而且现在形势严峻，时间紧迫，作战部越早得到情报，对作战越有益。

将军还说到，他的军队最缺的就是武器，尤其是用来摧毁敌人碉堡的大炮。此外，他们的弹药数量也有限，而且枪支并不统一，虽然都是来复枪，但牌子不同，型号各异。这为他们的作战带来了困难。他很想用美国的来福枪重新武装自己的队伍，那样队伍实力就会大增，战胜西班牙军队的把握也就更大了。

陪我回美国的三名军官分别是：克拉左将军，一位杰出的指挥官；赫蓝德兹上校；威塔医生，他对岛上的疾病和热带雨林了如指掌。此外，加西亚将军还安排两名水手与我们一同上路，这二人十分熟悉古巴北海岸的情况。如果美国真的为古巴提供他们需要的武器，那么在武器运送的途中，他们能发挥相当重要的作用。

我还可以继续考察下去吗？我可以提更多方面的问题吗？在过去的九天中，我走过了许多状况各异的地形，多么希望能有机会让我更加仔细地观察这奇异的环境，以便给总统一个满意的答案。但是，当将军问："你还有什么问题吗？"我的回答干脆而简洁："没有，将军！"为什么不这样回答呢？加西亚将军用他敏锐的眼光、机智的头脑、快速的适应能力，使我省去了数月的劳苦奔波。况且，在现有的情形下，我们国家可以获得更详细的情报，和古巴军队掌握的情报一样。这等于通过他们掌握了敌人的情况。

接下来的两个小时里，我得到了非正式的热情招待。正式的宴会在五点钟开始，宴会结束后，我得知随我回美国的人已经到了门口。当我见到他们时大吃一惊，原来队伍中没有我来时路上的向导和同伴。我问格瓦希奥以及其他人在哪里。经过一番解释，我才明白，原来格瓦希奥是想和我一起回美国，但加西亚将军反对，因为南方海岸的战争需格瓦希奥这样的人才，而我是从北方返回美国。我衷心感谢格瓦希奥和他的船员们以及瑞奥斯将军的两百名骑兵，并向加西亚将军表达了自己的这种感激之情。我用拉丁式的拥抱和将军告别，然后上马离开。当我们的马奔向北方的时候，我听到了身后的欢呼声。

最终，我把信送到了加西亚将军的手中！

虽然在送信的过程中遭遇了很多危险，与我回到美国的行程相比，也要重要很多。在来的路上没有几次战斗，一路上我有很棒的向导，并且得到了无微不至的照顾，我还欣赏了这美丽的国家。但此时，战争已经爆发，西班牙人的警惕性更高了，他们的士兵到处巡逻，不放过每一处海岸，甚至每一个海湾和小港都派兵驻守。他们早把大炮准备好了，谁破坏战争规则，谁就是被炮轰的人。只要我在他们的领域被发现，不管我如何解释和隐瞒，他们都会认为我是间谍。一旦被发现，等待我的就是死亡。

面对波涛汹涌的大海和阴霾的天空，我认为成功从来不是一帆风顺的。但是，我们必须努力前行，并取得成功，否则我的任务就没有圆满地完成。任务的圆满完成，可能在很大程度上取决于我们对西班牙战争的胜利。

在返程的路上，同伴们也不轻松，恐惧和忧虑同样弥漫在他们心中。于是，我们小心翼翼地穿越过了古巴，一路向北前行，来到了考托内河码头。这里是水运航道的枢纽，但处在西班牙的控制之下。当我们进入呈瓶状的马纳提海港后，才发现对面的海岸上耸立着敌人的碉堡，里面的大炮已经对准了河口。一旦西班牙人发现了我们，我们必死无疑！但是，在此时，我们只能选择前进，勇气反而成了我们的救星。西班牙人怎么会想到像我们这样肩负重任的“敌人”会有胆量在这里上船呢?

我们乘坐的船，体积仅为104立方英尺，是一条名副其实的小船。船上没有帆，我们就把麻布袋做成帆。靠着每天微量的熟牛肉和水，我们扬帆航行了150英里，来到了新普罗维登斯的拿骚岛。想一想，我们乘坐一只设备简陋的小船航行在西班牙人的海域里，而他们装备良好的驱逐舰四周巡逻。对我们而言，形势多么危险啊！但是，完成任务的使命感让我们勇往直前。因为只有这样，我们才可能完成任务。

船太小了，它无法承载6个人走更远的路程，于是威塔医生返回了巴亚莫。剩下5个人要冒着西班牙人的强大火力，扬起用麻布袋做的帆，凭借机智，驾驶着小船，在西班牙巡洋舰的威慑下闯出一条生路！

正当我们准备出发时，一场暴风雨突然降临，海上巨浪滔天。看着变得非常坏的天气，我们不敢贸然航行。但问题是，我们原地不动也同样危险。因为天空中挂着一轮满月，只不过暂时被乌云遮住了。只要云层散去，我们就会被敌人发现。但不管怎样，我坚信命运掌握在自己的手中。

11点时，我们登上船。因为少了一个人，小船行进得还算顺利。天空乌云密布，遮住了月亮。五个人当中，四人划桨，一人掌舵。西班牙人的要塞逐渐远去，或者这样说更确切，那就是要塞里的敌人没有发现我们。虽然前行得很艰难，但总算没有听到大炮的轰鸣声和机枪的扫射声。我们的小船摇摇晃晃，像一只蛋壳，好几次差一点就要翻船了。幸亏我们的水手谙熟水性，麻布袋做的帆也经受住了考验，于是，我们继续航行。

我们从一个浪头划到下一个浪头，过度的疲劳，单调的航程，让我们昏昏欲睡。不久，一个巨浪袭来，差点把小船掀翻，船里也灌满了水。从这之后，大家的睡意都无影无踪了。因为我们需要不停地向外舀水。漫长的黑夜终于过去了。我们全身都湿透了，而且又累又饿。但是，当看到太阳穿越薄雾出现在地平线上，我们都非常兴奋。

“快来看，先生！”舵手指着一艘蒸汽船大声喊道。

大家的心立刻提到了嗓子眼！难道那是一艘西班牙战舰？如果真是这样，那我们就危在旦夕。

"Dos vapores, tres vapores, Caramba! doce vapores!"舵手用西班牙语喊着，其他同伴也纷纷应和。真的是西班牙战舰吗?

不，最后证实这是桑普森海军上将的战舰，正向东航行，去抗击西班牙的战舰!

这时，我们的心才回到了原地。

天气炎热，可是我们不能休息，要不停地向外舀船里的水。人人都很紧张，睡意全无。虽然我们看见了美国的战舰，但不能保证西班牙炮艇不会出现，因为可能会有逃过美国军舰警戒的"漏网之鱼"。一旦有，他们就会追上我们，然后将我们抓住。又到傍晚了，我们再也没有力气了，似乎随时都能倒下。可是，我们依然不敢休息。随着夜晚的到来，海风刮了起来，风力逐渐增加，波涛越来越大。为了使小船不至于倾覆，舀水的工程又开始了。第二天早上（5月7日）大概10点钟的时候，情况才有所好转。我们来到了巴哈马群岛安德罗斯岛南端一个叫克里基茨的地方，在这里登上了陆地，并短暂休息了一下。

下午，在13个黑人船员的帮助下，我们里里外外地检查和修理了小船。这些黑人操着一口怪异的方言，我们根本不知道他们在说什么。不过，通过手势，总算明白了彼此的意思。我们快速地装好了物品，带了一些猪肉罐头和一把手风琴。尽管我的体力已经严重透支，可是却难以入睡，因为手风琴不时发出刺耳的尖叫声。我再也不想听手风琴的声音了!

5月8日下午，我们绕过新普罗维登斯岛，在向西航行之时，被检疫官员抓起来了，他们将我们关在豪格岛上，理由是，我们染上了古巴黄热病。

但是第二天，我给美国总领事馆领事麦克莱恩先生捎去了一个口信。在他的安排下，5月10日，我们被释放了。5月11日，这只“无畏号”小船离开了码头，又再次航行了。

小船到达佛罗里达海域时，我们的运气似乎很差，一整天都没有风，小船无法前进。直到晚上，海面上才有了些微风。5月13日的早上，我们已经来到了基维斯特。

晚上，我们乘火车来到塔姆帕，然后换乘前往华盛顿的火车，并在预定的时间到达。接着，我向战事秘书罗塞尔·阿尔杰汇报了工作。我汇报完毕后，他让我带着加西亚将军派过来的人向迈尔斯将军汇报。听完汇报后，迈尔斯将军给作战部写了一封信：“我推荐美国第十九步兵团的一等中尉安德鲁·罗文为骑兵团上校副官。他出色地完成了古巴之行，把信送给了古巴起义军领袖加西亚将军，并为美国政府带回来了极其宝贵的情报。这是一项十分艰巨的任务，我认为罗文中尉在完成任务的过程中显示出了勇往直前和沉着冷静的高贵品质。他的精神将永载史册！”

大概在我返回的第二天，我陪同迈尔斯将军参加了一次内阁会议。在会议结束时，我收到了麦金莱总统的贺信，他感谢我将他的愿望传达给了加西亚将军，并对我的工作给予了高度的评价。

在信的最后，他写道：“你勇敢地完成了任务！”虽然这件事对我而言是第一次，但我只是完成了一次普通的任务，一个军人应该完成的任务："不要考虑理由，只要服从命令。"

我最终把信送给了加西亚将军。

"Where," asked President McKinley of Colonel Arthur Wagner, head of the Bureau of Military Intelligence, "where can I find a man who will carry a message to Garcia?"

The reply was prompt. "There is a young officer here in Washington; a lieutenant named Rowan, who will carry it for you!"

"Send him!" was the President's order.

The United States faced a war with Spain. The President was anxious for information. He realized that success meant that the soldiers of the republic must cooperate with the insurgent forces of Cuba. He understood that it was essential to know how many Spanish troops there were on the island, their quality and condition, their morale, the character of their officers, especially those of the high command; the state of the roads in all seasons; the sanitary situation in both the Spanish and insurgent armies and the country in general; how well both sides were armed and what the Cuban forces would need in order to harass the enemy while American battalions were being mobilized; the topography of the country and many other important facts. Small wonder that the command, "Send him!" was equally as prompt as the answer to his question respecting the individual who would carry the message to Garcia.

It was perhaps an hour later, at noon, when Colonel Wagner came to me to ask me to meet him at the Army and Navy Club for lunch at one o'clock. As we were eating, the colonel—who had, by the way, a reputation for being an inveterate joker—asked me: "When does the next boat leave for Jamaica?" Thinking he was making an effort to perpetrate one of his pleasantries, and determined to thwart him, if possible, I excused myself for a minute or so and when I had returned informed him that the "Adirondack", of the Atlas Line, a British boat, would sail from New York the next day at noon.

"Can you take that boat?" snapped the colonel.

Notwithstanding that I still believed the colonel was joking I replied in the affirmative. "Then," said my superior, "get ready to take it."

"Young man," he continued, "you have been selected by the President to communicate with—or rather, to carry a message to—General Garcia, who will be found somewhere in the eastern part of Cuba. Your problem will be to secure from him information of a military character, bring it down to date and arrange it on a working basis. Your message to him will be in the nature of a series of inquiries from the President. Written communication, further than is necessary to identify you, will be avoided. History has furnished us with the record of too many tragedies to warrant taking risks. Nathan Hale of the Continental Army, and Lieutenant Richey in the War with Mexico were both caught with dispatches; both were put to death and in the case of the latter the plans for Scott's invasion of Vera Cruz was divulged to the enemy. There must be no failure on your part; there must be no errors made in this case."

By this time I was fully alive to the fact that Colonel Wagner was not joking. "Means will be found," he continued, "to identify you in Jamaica, where there is a Cuban junta. The rest depends on you. You require no further instructions than those I will now give you." Which he did, they being essentially as outlined in the opening paragraphs. "You will need the afternoon for preparation. Quartermaster—General Humphreys will see that you are put ashore at Kingston. After that, providing the United States declares war on Spain, further instructions will be based on cables received from you. Otherwise everything will be silence. You must plan and act for yourself. The task is yours and yours only. You must get a message to Garcia. Your train leaves at midnight. Goodbye and good luck!" We shook hands.

As Colonel Wagner released mine he repeated: "Get that message to Garcia!" Hastily, as I set about to make my preparations, I considered my situation. My duty was, as I understood it, complicated by the fact that a state of war did not exist, nor would it exist at the time of my departure; possibly not until after my arrival in Jamaica. A false step might bring about a condition that a lifetime of statement would never explain. Should war be declared my mission would be simplified, although its dangers would not be lessened. In instances of this kind, where one's reputation, as well as his life, is at stake, it is usual to ask for written instructions. In military service the life of the man is at the disposal of his country, but his reputation is his own and it ought not to be placed in the hands of anyone with power to destroy it, either by neglect or otherwise. But in this case it never occurred to me to ask for written instructions; my sole thought was that I was charged with a message to Garcia and to get from him certain information and that I was going to do it. Whether Colonel Wagner ever placed on file in the office of the adjutant general the substance of our conversation I do not know. At this late day it matters little.

My train left Washington at 12:01 a.m., and I have a recollection of thinking of an old superstition about starting on a journey on Friday. It was Saturday when the train departed, but it was Friday when I left the club. I assumed the Fates would decide that I had left on Friday. But I soon forgot that in my mental discussion of other matters and did not recall it until some time afterward and then it mattered nothing, for my mission had been completed.

The "Adirondack" left on time and the voyage was without special incident. I held myself aloof from the other passengers and learned only from a traveling companion, an electrical engineer, what was going on. He conveyed to me the cheerful information that because of my keeping away from them and giving no one any information as to my business, a bunch of convivial spirits had conferred on me the title of "the bunco steerer".

It was when the ship entered Cuban waters that I first realized danger. I had but one incriminating paper, a letter from the State Department to officials in Jamaica saying that I was what I might represent myself to be. But if war had been declared before the Adirondack entered Cuban waters she would have been liable to search by Spain, under the rules of international law. As I was contraband and the bearer of contraband I could have been seized as a prisoner of war and taken aboard any Spanish ship, while the British boat, after compliance with specified preliminaries, could have been sunk, despite the fact that she left a peaceful port under a neutral flag, bound for a neutral port, prior to a declaration of war.

Recalling this state of affairs, I hid this paper in the life preserver in my stateroom and it was with great relief I saw the cape astern. By nine the next morning I had landed and was a guest of Jamaica. I was soon in touch with Mr. Lay, head of the Cuban junta, and with him and his aids planning to get to Garcia as soon as possible. I had left Washington April 8-9. April

20 the cables announced that the United States had given Spain until the 23 to agree to surrender Cuba to the Cubans and to withdraw her armed forces from the island and her navy from its waters. I had in cypher cabled my arrival and on April 23 a reply in code came: "Join Garcia as soon as possible!"

In a few minutes after its receipt I was at headquarters of the junta, where I was expected. There were a number of exiled Cubans present whom I had not met before and we were conversing on general topics when a carriage drove up.

"It is time!" some one exclaimed in Spanish.

Following which, without further discussion, I was led to the vehicle and took a seat inside.

Then began one of the strangest rides ever taken by a soldier on duty or off. My driver proved to be the most taciturn of Jehus. He spoke not to me, nor heeded me when I spoke to him. The instant I was shut in he started through the maze of Kingston's streets at a furious pace. On and on he drove, never slackening speed, and soon we had passed the suburbs and were beyond all habitations. I knocked, yes, kicked, but he gave no heed.

He seemed to understand that I was carrying a message to Garcia and that it was his part to get me over the first "leg" of the journey as speedily as possible. So, after several futile efforts to make him listen to me, I decided to let matters take their course and settled back in my seat.

Four miles further, through a dense growth of tropical trees, we flew along the broad and level Spanish Town road, until at the edge of the jungle we halted, the door of the cab was opened, a strange face appeared, and I was invited to transfer to another carriage that was waiting. But the strangeness of it all! The order in which everything appeared to be arranged! Not an unnecessary word was indulged in, not a second of time was wasted.

A minute later and again I was on my way. The second driver, like the first, was dumb. He declined all efforts made to get him in conversation, contenting himself by putting his horses to as swift a pace as possible, so on we went through Spanish Town and up the valley of the Cobre river to the backbone of the island where the road runs down to the ultramarine waters of the Caribbean at St. Ann's Bay.

Still not a word from my driver, although I repeatedly endeavored to get him to talk to me. Not a sound, not a sign that he understood me: just a race along a splendid road, breathing more freely as the altitude increased, until as the sun set we drew up beside a railway station. But what is this mass of ebony rolling down the slope of the cut toward me? Had the Spanish authorities anticipated me and placed Jamaica officers on my trail?

I was uneasy for a moment as this apparition came in sight, but relief came when an old negro hobbled to the carriage and shoved through the door a deliciously fried chicken and two bottles of Bass' ale, at the same time letting loose a volley of dialect, which, as I was able to catch a word here and there, I understood was highly complimentary to me for helping Cuba gain her freedom and giving me to understand that he was "doing his bit" with me.

But my driver stood not on ceremony, nor was he interested in either chicken or conversation. In a trice a new pair of horses was relayed on and away we went. My Jehu plying his whip vigorously. I had only time enough to thank the old Negro by shouting: "Goodbye, Uncle!" In another minute we had left him and were racing through the darkness at breakneck speed. Although I fully comprehended the gravity and importance of the errand in which I was engaged, I lost sight of it for the time in my admiration of the tropical forests. These wear their beauty at night as well as by day. The difference is that while during the sunlight it is the vegetable world that is in perennial bloom, at night it is the insect world in its flight that excites attention. Hardly had the short twilight changed to utter darkness when the glowworms turned on their phosphorescent lights and flooded the woods with their weird beauties. These magnificent fireflies illuminated with their incandescence the forest I was traversing until it resembled a veritable fairyland.

But even such wonders as these are forgotten in the recollection of duty to be performed. We still coursed onward at a speed that was limited only by the physical abilities of the horses, when suddenly a shrill whistle sounded from the jungle! My carriage stopped. Men appeared as if they had sprung from the ground. I was surrounded by a party of men armed to the teeth. I had no fear of being intercepted on British soil by Spanish soldiers, but these abrupt halts were getting on my nerves, because action by the Jamaica authorities would mean the failure of the mission, and if the Jamaica authorities had been notified that I was violating the neutrality of the island I would not be allowed to proceed. What if these men were English soldiers! But my feelings were soon relieved. A whispered parley and we were away again!

In about an hour we halted in front of a house outlined by feeble lights within. Supper waited. The junta manifestly believed in liberal feeding. The first thing offered me was a glass of Jamaica rum. I do not recall that I was tired, although we had traveled about seventy miles in approximately nine hours with two relays, but I do know that the rum was welcome. Following came introductions. From an adjoining room came a tall, wiry, determined-looking man, with a fierce moustache, one of his hands minus a thumb; a man to tie to in an emergency, to trust at any time. His eyes were honest, loyal eyes that mirrored a noble soul. He was a Peninsula Spaniard who had gone to Cuba, at Santiago had quarreled with the rule of Old Spain, hence the missing thumb and exile. He was Gervacio Sabio and he was charged with seeing that I was guided to General Garcia for the delivery of my message. The others were the men employed to get me out of Jamaica—seven miles remaining to be traveled—with one exception, one man was to be my "assistente", or orderly.

Following a rest of an hour, we proceeded. Half an hour's travel from the hut we were again halted by whistle signals. We alighted and entered a cane field through which we tramped in silence for about a mile until we came to a cocoanut grove bordering a plaything of a bay. Fifty yards off shore a small fishing boat rocked softly on the water.

Suddenly a light flashed aboard the little craft. It must have been a time signal, for our arrival had been noiseless. Gervacio, apparently satisfied with the alertness of the crew, answered it. Following some conversation during which I thanked the agents of the junta, I climbed on the back of one of the boat's crew who had waded ashore and was carried to the boat. I had completed the first part of the journey to Garcia.

Once aboard the boat I noted that it was partially filled with boulders intended for ballast. Oblong bundles indicated cargo, but not sufficient to impede progress. But with Gervacio as skipper, the crew of two men, my assistente and myself, the boulders and the bundles, there was little room for comfort. I indicated to Gervacio my desire to get beyond the three-mile limit as soon as possible, as I did not want to impose upon the hospitality of

Great Britain longer than necessary. He replied that the boat would have to be rowed beyond the headlands, as there was not sufficient wind in the small bay to fill her sails. We were soon outside the cape, however, our sails caught the breeze and the second stretch of the trip to the strife-torn objective was begun.

I have no hesitation in saying that there were some anxious moments for me following our departure. My reputation was at stake if I should be caught within the three-mile limit off the Jamaica coast. My life would be at stake if I should be caught within three miles of the Cuban coast. My only friends were the crew and the Caribbean Sea.

One hundred miles to the north lay the shores of Cuba, patrolled by Spanish "lanchas", light-draft vessels armed with pivot guns of small caliber, and machine guns, their crews provided with Mauser rifles, far superior—as I afterward learned—to anything we had aboard, as motley a collection of small arms as could be picked up anywhere. In the event of an encounter with one of these "lanchas" there was little to hope for. But I must succeed, I must find Garcia and deliver my message!

Our plan of action was to keep outside the Cuban three-mile limit until after sunset, then to sail or row in rapidly, draw behind some friendly coral reef and wait until morning. If we were caught, as we carried no papers, we would probably be sunk and no questions asked. Boulder-laden craft go to the bottom quickly and floating bodies tell no tales to those who find them. It was now early morning, the air was deliciously cool and, wearied with my journey thus far I was about to seek some rest in sleep when suddenly Gervacio gave an exclamation that brought us all to our feet. A few miles away one of the dreaded lanchas was bearing directly toward us.

A sharp command in Spanish and the crew dropped the sail. Another and all save Gervacio, who was at the helm, were below the gunwale, and he was lounging over the tiller, keeping the boat's nose parallel with the Jamaica shore. "He may think I am a 'lone fisherman' from Jamaica and go by us." said the cool-headed steersman. So it proved. When within hailing distance the pert young commander of the lancha cried in Spanish: "Catching anything?"

To which my guide responded, also in Spanish: "No, the miserable fish are not biting this morning!"

If only that midshipman, or whatever his rank, had been wise enough to lay alongside, he surely would have "caught something", and this story would never have been written. When he had passed us and was some distance away, Gervacio ordered sail hoisted again and turning to me, remarked: "If the Senor is tired and wants sleep, he can now indulge himself, for I think the danger is past."

If anything occurred during the next six hours, it left me undisturbed. In fact, I believed that nothing except the broiling heat of the tropical sun could have drawn me from my rocky mattress. But it did for the Cubans, who were quite proud of their English greeted me with: "Buenos dias, Meester Rowan!" The sun shone brilliantly all day. Jamaica was all aglow, like some mighty jewel in a setting of emerald. The turquoise sky was cloudless and to the south the green slopes of the island were blocked off in large squares, showing to great advantage the light verdancy of the cane fields alternating with the deeper hue of the forests. It was a splendid and a magnificent picture. But northward all was gloom. An immense bank of clouds enshrouded Cuba and, watch as keenly as we might, we saw no sign of their lifting. But the wind held true and even increased in volume during the hours. We were making good progress and Gervacio at the tiller was happy, joking with the crew and smoking like a "fumarole" .

About four o'clock in the afternoon the clouds broke away and the Sierra Maestra, the master mountain range of the island, stood in the golden sunshine in all its beauteous majesty. It was like drawing the curtain aside and placing on view a matchless picture by an artist monarch. Here were color, mass, mountain, land and sea blended in one splendid ensemble, the like of which is found nowhere else, for there is no place on earth where a mountain height of 8000 feet, its summits clothed in verdure and its great battlements extending for hundreds of miles!

But my admiration was short lived. Gervacio broke the spell when he began taking in sail. To my question he replied: "We are closer than I thought. We are in the war zone of the lanchas, high seas or no high seas. We must stand well out and use the open water for all it is worth. To go closer and run the risk of being seen by the enemy is merely to run an unnecessary risk."

Hastily we overhauled the arsenal. I carried only a Smith & Wesson revolver, so I was assigned a frightful looking rifle. I might have been able to fire it once, but I doubt if it would have been of further service. The crew and my assistant were provided with the same formidable weapons, while the pilot, who from his seat looked after the jib, the only sail set, drew close to him the other weapons. The real serious part of my mission was now at hand. Hitherto everything had been easy and comparatively safe. Now danger menaced. Grave danger. Capture meant death and my failure to carry my message to Garcia.

We were probably twenty five miles from the coast, although it seemed but a span away. It was not until nearly midnight that the jib-sheet was let go and the crew began sounding the shallow water with their oars. Then a timely roller gave us a last lift and with a mighty effort shoved us into the waters of a hidden peaceful bay. We anchored in the darkness fifty yards off shore. I suggested that we land at once, but Gervacio replied: "We have enemies both ashore and afloat,

Senor; it is better that we stay where we are. Should any lancha endeavor to pry us out she would likely land on the submerged coral reef we have crossed and we can get ashore, and from the obscurity of the grape entanglements we can play the game."

The tropical haze which ever hangs mist like at the meeting of the sea and sky in low altitudes began to lift slowly, disclosing a mass of grape, mangrove thickets and thorn-set trees, reaching almost to the edge of the water. It was difficult to perceive objects with distinctness, but as if declining to puzzle us further as to the nature of our surroundings, the sun rose gloriously over El Turquino, the highest point in all Cuba. In an instant everything had changed, the mist had vanished, the darkness of the low-lying thicket against the mountain wall had been dissipated, the gray of the water breaking against the shore had been transformed as if by magic to a marvelous green. It was one splendid triumph of light over darkness.

Already the crew were busy transferring luggage ashore. Noting me standing mute and seemingly dazed, for I was thinking of the lines by a poet who must have had a similar scene in mind when he wrote: "Night's candles are burnt out and jocund day stands tip-toe on the misty mountain tops." Gervacio said in a low tone to me: "El Turquino, Senor."

As I stood there drinking in the glory of that marvellous morning, little did I dream that I was standing within a stone's throw, almost, of what was soon to be the watery sepulchre of the mighty"Colon", a great battleship, then first in her class and bearing the name of the greatest of all admirals, Christopher Columbus, the discoverer of America, this great ship having already been selected by the Fates to be destroyed by our own warships in the sea fight off Santiago.

But my reveries were soon ended. The freight was landed, I was carried ashore, the boat dragged to a small estuary, overturned and hidden in the jungle. By this time a number of ragged Cubans had assembled at our landing place. Where they came from, or how they knew that our party was a friendly one, were problems too deep for me. Signals of some sort had doubtless been exchanged and they had come to act as burden-bearers. Some of them had seen service, some of them bore the marks made by Mauser bullets.

Our landing place seemed to be a junction of paths running in all directions away from the coast and into the thicket. Off to the west, seemingly about a mile away, little columns of smoke were rising through the vegetation. I learned that this smoke was from a "salina", or pan where salt was being made for the refugee Cubans who had hidden in these mountains after fleeing from the dreaded concentration camps.

The second "leg" of the journey was completed.

Hitherto there had been danger; from this time on there would be more. Spanish troops mercilessly hunted down Cubans and small mercy was shown by the forces directed by Weyler, the "butcher", to men found in arms, or outside the concentration camps, even though they might be unarmed. The remainder of the journey to Garcia was fraught with many dangers and I knew it, but this was no time to consider them; I must be on my way!

The topography of the country was simple enough; a level strip of land extending a mile or so inland toward the north, covered with jungle. Man's handiwork had been confined to cutting paths, and the network could be threaded only by the Cubans reared in this labyrinth. The heat soon became oppressive and caused me to envy my companions, none of whom were burdened by superfluous clothing.

Soon we were on the march, screened from the sea and the mountains, and indeed, from each other, by the denseness of the foliage, the twists and turns of the trail and the torrid haze that soon settled over everything. The jungle was converted into a miniature inferno by the sun, although we could not see it through the verdure. But as we left the coast and approached the foothills the jungle began to give way to a larger and less dense growth. We soon reached a clearing where we found a few bearing coconut trees. The water fresh and cool, drawn from the nuts, was elixir to our parched throats. But not long did we tarry in this pleasant spot. A march of miles lay before us and a climb up steep mountain slopes to another hidden clearing must be made before nightfall. Soon we had entered the true tropical forest. Here traveling was somewhat easier, for a current of air, hardly perceptible, but a current of air nevertheless, made breathing less of a task and, by far, more refreshing.

Through this forest runs the "Royal Road" from Portillo to Santiago de Cuba. As we neared this highway I noted my companions one by one disappearing in the jungle. I was soon left alone with Gervacio. Turning to him to ask a question I saw him place a finger on his lips, mutely sign to me to have my rifle and revolver in readiness and then he too vanished amid the tropical growth.

I was not long in ascertaining the reason for this strange conduct. The jingle of horses' trappings, the rattling of the short sabers carried by Spanish cavalry and occasionally a word of command, fell on my ear. But for the vigilance of those with me we should have walked out on the highway just in time to encounter a hostile force! I cocked my rifle and swung my Smith & Wesson into position for quick action and waited tensely for what was to follow. Every moment I expected to hear reports of firearms. But none came and one by one the men returned, Gervacio being among the last. "We scattered in order to deceive them in the event we had been discovered. We covered a considerable stretch of the road and had firing been commenced the enemy would have believed it an attack in force from ambush. It would have been a successful one too," Gervacio added with an expression of regret, "but duty first and—" here he smiled, " pleasure afterward."

Beside the trails along which insurgent parties usually passed, it was the custom to build fires and bury sweet potatoes in the ashes. There they roasted until a hungry party should pass. We came upon one of these fires during the afternoon. A baked sweet potato was passed out to each of the party, the fire covered again and the march resumed.

As we ate our sweet potatoes I thought of Marion and his men in the days of the revolution, who fought their battles on a like diet, and through my mind flashed the idea that as Marion and his men had fought to victory, so also would these Cubans, who were inspired by a desire for liberty similar to that actuating the patriot fathers of my own country, and it was with a feeling of pride that I recalled that my mission was to aid these people in their efforts by communicating with their general and making it possible for the soldiers of my nation to do battle in their behalf.

Arriving at the end of the journey for the day, I observed a number of men in a dress strange to me.

"Who are these?" I inquired.

"They are deserters from the army of Spain, Senor," replied Gervacio. "They have fled from Manzanillo and they say that lack of food and harsh treatment by their officers were the reasons for their leaving."

Now a deserter is sometimes of value, but here in this wilderness I would have preferred their room to their company. Who could say that one or more of them might not leave camp at any time and warn the Spanish officials that an American was crossing Cuba, evidently bound for the camp of General Garcia? Would not the enemy make every effort to thwart him in his mission? So I said to Gervacio: "Question these men closely and see that they do not leave camp during our stay!"

"Si, Senor!" was the reply.

Well for me and the success of my errand that I had give out such instruction. My thought that one or more deserters might leave to apprise the Spanish commander of my presence proved to be the correct one. Although it is not fair to presume that any knew my mission, my being there was sufficient to arouse the suspicions of two who proved to be spies and also nearly resulted in my assassination. These two determined to leave camp that night and plunge through the thickets to the Spanish lines with the information that an "officer Americano" was being escorted across Cuba.

I was awakened some time after midnight by the challenge of a sentinel, followed by a shot, and almost instantly a shadowy form appeared close by my hammock. I sprang up and out on the opposite side just as another form appeared and in less time than it takes to write it the first one had fallen as the result of a blow from a machete, which cut through the bones of his right shoulder to the lung. The wretch lived long enough to tell us that it was agreed if his comrade failed to get out of camp, he should kill me and prevent the carrying out of whatever project I was engaged in. The sentinel shot and killed his comrade.

Horses and saddles were not available until late next day, at an hour that made it impossible to proceed. I chafed at the delay, but it could not be helped. Saddles were harder to secure than horses. I was somewhat impatient and asked Gervacio why we could not proceed without saddles.

"General Garcia is besieging Bayamo, in Central Cuba, Senor," was his reply, " and we shall have to travel a considerable distance in order to reach him." This was the reason for the search for "monturas", the saddles and trappings. One looked at the steed assigned me and my admiration for the wisdom of my guide mounted rapidly and increased noticeably during the four days' ride. Had I ridden that skeleton without a saddle it would have meant exquisite torture. However, I will say for the horse, that with his "montura" he proved a mettlesome beast, far superior to many a well-fed horse of the plains of America.

Our trail followed the backbone of the ridge for some distance after leaving camp. One unaccustomed to these trails must surely have been driven desperate by the perplexity of the wilderness, but our guides seemed to be as familiar with the tortuous windings as they would have been on a broad high road.

Shortly after we had left the divide and had begun the descent of the eastern slope we were greeted by a motley assembly of children and an old man whose white hair streamed down his shoulders. The column halted, a few words passed between the patriarch and Gervacio, and then the forest rang with "Viva!" for the United States, for Cuba and the "Delegado Americano". It was a touching incident. How they had learned of my approach I never knew; but news travels fast in the jungle and my arrival had made one old man and a crowd of little children happier.

At Yara, where the river leaves the foothills we camped that night, it was brought to me that we were in a zone where danger lurked. "Trincheras" or trenches had been built to defend the gorge should the Spanish columns march out from Manzanillo. Yara is a great name in Cuban history, for from the town of Yara came the first cry for "liberty" in the "Ten Years' War" of 1868—1878. I was asked to swing my hammock behind the trinchera, which, by the way, was not a trench at all, but a breast-high wall of stones, and I noticed that a guard, recruited from some unknown source, was posted and kept on duty all night. Gervacio intended taking no chances on my mission being a failure.

Next morning we began the ascent of the spur projecting northward from the Sierra Maestra, forming the east bank of the river. Our course lay across the eroded ridges. Danger lurked in the lowlands. There was the possibility of ambuscade, fire and the chance of being cut off by some mobile party of Spaniards.

Here began a series of ups and downs across the streams with vertical banks. In my career I have seen much cruelty to animals, but never anything to equal this. To get the poor horses down to the bottom of these gulches and out again involved forms of punishment beyond belief. But there was no help for it; the message to Garcia must be delivered, and in war what are the sufferings of a few horses when the freedom of hundreds of thousands of human beings is at stake? I felt sorry for the brutes, but this was no time for sentiment.

It was with great relief that after the hardest day of riding I had ever experienced we halted at a hut in the midst of corn patches near the edges of the forest, at Jibaro. A freshly killed beef was hanging to the rafters, while the cook in the open was busy preparing a meal for the "Delegado Americano". My coming had been heralded and my feast was to consist of fresh beef and cassava bread.

Hardly had I finished my generous meal when a great commotion was heard, voices and the clatter of horses' hoofs at the edge of the forest. Colonel Castillo of the staff of General Rios had arrived. He welcomed me in the name of his chief, who was due to arrive in the morning, with all the grace of a trained staff officer; then mounting his steed with an athletic spring, put the spurs to his mount in frenzied fashion and was off, as he came, like a flash.

His welcome assured me that I was making headway under a skillful guide.

General Rios came next morning and with him Colonel Castillo,who presented me with a Panama hat "made in Cuba".

General Rios was "the general of the coasts". He was very dark, evidently of Indian and Spanish blood, with springy, athletic step. No Spanish column ever made a sortie in his district and found him unprepared. His sources of information and his intuition were uncanny. It was no small task to move hiding families and provide for their maintenance, but he did it and, as may be supposed, advance information of enemy movements was imperative. The Spanish methods were to enter the forests, scour them and, in default of prey, lay the districts in waste. Meanwhile General Rios would conduct matters in guerilla fashion and his forces were continuously taking pot shots at the Spanish columns, sometimes doing terrible execution.

General Rios added two hundred cavalrymen to my escort. As we marched single file we would have presented a formidable appearance had there been anyone to see us. I could not help observing that we were being led with remarkable skill and speed. We had entered the forest again and were hiding in the evergreen dress of the Sierra Maestra. The trail was comparatively level, but crossed at intervals by water courses with steep banks. The paths were so narrow we were constantly running afoul of tree trunks, barking our shins and dislodging the impedimenta from the backs of our horses. Still the guide held to a steady gait that caused me to marvel. My usual position was near the center of the column, but I wanted to be near this centaur who was in the lead and at the next water course crossing I rode forward to observe him. He was a coal black Negro, Dionisito Lopez, a lieutenant in the Cuban army. He could trace a course through this trackless forest, through the tangled growth, as fast as he could ride. His skill with a machete was amazing. He carved a way for us through the jungle. Networks of vines fell before his steady strokes right and left; closed spaces became openings; the man appeared tireless.

The night of April 30 brought us to the Rio Buey, an affluent of the Bayamo River, and about twenty miles from the city of Bayamo. Our hammocks had scarcely been swung when Gervacio appeared, his face aglow with satisfaction. "He is there, Senor! General Garcia is in Bayamo and the Spaniards are in retreat down the Cauto river. Their rear-guard is at Cauto-El-Embarcadero!"

So eager was I to get in communication with Garcia that I proposed a night ride, but after a conference it was decided that nothing would be gained.

May-day, 1898, is "Dewey Day" in our calendar. As I was sleeping in the forests of Cuba, the great admiral was feeling his way past the guns of Corregidor into Manila Bay to destroy the Spanish fleet. While I was on my way to Garcia that day he had sunk the Spanish ships and with his guns was menacing the capital of the Philippines.

Early that morning we were on our way. Terrace by terrace we descended the slope leading to the plain of Bayamo. This great stretch of country, laid waste for years, was now as if man had never been. At the black remnant of the hacienda of Candalaria, mute evidence of Spanish methods of warfare, we passed into the plain. We had ridden more than one hundred miles through a wilderness with hardly a habitation to show that man had ever lived in one of Nature's most favored spots across a tropical garden gone to weeds. Through grass so high that our column was hidden from sight, through burning sun and blistering heat, we traveled, but all our discomforts were forgotten in the thought that our destination was at hand; our mission nearly ended. Even our jaded horses seemed to share in our anticipation and eagerness.

We struck the royal road to Manzanillo-Bayamo and encountered joyous human beings in rags and tatters, all hurrying toward the town. The chatter of these happy groups reminded me of the parrots that had shrieked at our passage through the jungles. They were going back to the homes from which they had been driven.

It was but a short ride from Paralejo to the banks of the eastern side of the river to the town, once a city of 30 000, now a mere village of perhaps 2 000. It was surrounded by a row of blockhouses the Spaniards had built on both sides of the stream. These little forts were the first objects to be seen and their prominence was emphasized by the flames and smoke still rising as we came into view. The Cubans had set them on fire when they entered the former metropolis of this once flourishing valley.

We soon lined up on the bank, and after Gervacio and Lopez had talked to the guards, we proceeded. We halted in mid-stream to allow our horses to drink and to store up a little energy for our final dash into the presence of the officer in charge of Cuba's military destiny east of the Jucaro-Moron trocha. (I quote from the newspapers of the day: "The Cuban generals say the arrival of Lieutenant Rowan aroused the greatest enthusiasm throughout the Cuban army. There was no notice of his coming and the first seen of lieutenant Rowan was as he galloped up Calle Commercial, followed by the cuban guides who accompanied him.")

In a few minutes I was in the presence of General Garcia.

The long and toilsome journey with its many risks, its chances of failure, its chances for death, was over.

I had succeeded.

As we arrived in front of General Garcia's headquarters the Cuban flag was hanging lazily over the door from an inclined staff. The method of reaching the presence of a man to whom one is accredited in such circumstances was new to me.

We formed in line, dismounted together, and "stood to horse". Gervacio was known to the general, so he advanced to the door and was admitted. He returned in a short time with General Garcia, who greeted me cordially and asked me to enter with my "assistente". The general introduced me to his staff—all in clean white uniforms and wearing side arms—and explained that the delay was caused by the necessary scrutiny of my credentials from the Cuban junta at Jamaica, which Gervacio had delivered to him.

There is humor in everything. I had been described in letters from the junta as "a man of confidence". The translator had made me "a confidence man". Following breakfast we proceeded to business. I explained to General Garcia that my errand was purely military in its character, although I had left the United States with diplomatic credentials; that the President and the War Department desired the latest information respecting the military situation in Eastern Cuba. (Two other officers had been sent to Central and Western Cuba, but they were unable to reach their objectives.) Among matters it was imperative for the United States to know were the positions occupied by the Spanish troops, the condition and number of the Spanish forces; the character of their officers, especially of their commanding officers; the morale of the Spanish troops; the topography of the country, both local and general; communications, especially the conditions of the roads; in short, any information which would enable the American general staff to lay out a campaign. Last, but by no means least, General Garcia's suggestions as to a plan of campaign, joint or separate, between the Cuban armies and the forces of the United States. Also I informed him, my government would be glad to receive the same information respecting the Cuban forces, or as much as the general saw fit to give. If not incompatible with his plans, I would like to accompany the Cuban forces in the field in such capacity as he might see fit to assign me.

General Garcia meditated for a moment and then withdrew with all the members of his staff excepting Colonel Garcia, his son, who remained with me. About three o'clock the general returned and said he had decided to send three officers to the United States with me. These officers were men who had passed their lives in Cuba; were trained and tried; all knew the country, and in their particular capacities could answer all questions likely to be propounded. Were I to remain months in Cuba I might not be able to make so complete a report, and as time was the important element, the quicker the United States government got the information the better it would be for all concerned.

He went on to explain that his men needed arms, especially artillery, important in assaulting blockhouses. In ammunition he was very short, and the many rifles of varied calibre used made it difficult to get an ample supply. He thought it might be better to rearm his men with American rifles in order to simplify that question.

General Collazo, a noted figure; Colonel Hernandezand Doctor; Vieta, a valued relative who was familiar with the diseases of the island and the tropics generally, and two sailors, both familiar with the north coast, would go with us; they might be useful on the return expedition in case the United States should decide to furnish the supplies he wanted.

Could I proceed that day—hoy mismo? Could I ask more? Could I ask more? I had been continuously on the move for nine days in all kinds and conditions of terrain. I would have liked to have had a chance to look around me in these strange surroundings, but my answer was as prompt as his question. I simply replied: "Yes, sir!" Why not? General Garcia by his quick conception and speedy acceptance of conditions had saved me months of useless toil and had given my country the means of obtaining as minute information of the existing situation in the island as that possessed by the Cubans themselves; certainly as good as the enemy had.

For the next two hours I was the recipient of an informal reception. Then a final meal was served at five o'clock, and at its conclusion I was told that my escort was at the door. When I reached the street I was surprised not to see my former guide and companion in the column. I asked for Gervacio, and he and the others of the contingent from Jamaica came out. Gervacio wanted to go with me, but Garcia was adamant; all were needed for service on the south coast and I was to return by the north. I expressed to the general my appreciation for the services of Gervacio and his crew, and the column drafted from the fastnesses of Sierra Maestra. After a real Latin embrace I broke away and mounted. Three cheers rang out as we galloped northward.

I had delivered my message to Garcia!

My journey to General Garcia had been fraught with many dangers, but it was, compared with my trip back to the United States, by far the more important, an innocent ramble through a fair country. Going in there had been little to contend with, for the voyage from Jamaica had been on pleasant waters, while on the way to the Cuban commander I had been well guarded an well guided. But war had been declared and the Spanish were alert. Their soldiers patrolled every mile of shore, their boats every bay and inlet, the great guns of their forts stood ready to speak in no uncertain tones to anyone violating the rules of warfare. To all intents and purposes I was a spy within the enemy lines! Discovery meant death with one's face to the wall.

Nor had I thought of reckoning with the angry elements of sea and air, which soon were to convince me that success is not always a matter of fair sailing. But the effort must be made and it must be successful, otherwise my mission had been fruitless. On the happy termination of it might depend, in a large measure, the carrying to victory of the war.

My companions shared with me the apprehensions that naturally arose, so it was with great caution that we proceeded across Cuba, northward, going around the Spanish position at Cauto-El-Embarcadero, head of navigation on that river, at least for gunboats, until we came to the bottle-shaped harbor of Manati, where, on the side opposite, a great fort, bristling with guns, guarded the entrance. If only the Spanish soldiery had known of our presence! But perhaps the very audacity of our undertaking was our salvation. Who would have suspected that an enemy on a mission such as was ours, would select such a place from which to embark?

The boat in which we made the voyage was a cockleshell, "capacity 104 cubic feet". For sails we had gunnysacks, pieced together. For rations boiled beef and water. In this craft we were to sail, and we did sail, 150 miles due north to New Providence, Nassau Island. Think of putting to sea on hostile waters, patrolled by swift, well-armed lanchas, in a vessel like that! But "needs be when the devil drives!" It was our only method of fulfilling the full measure of duty.

It was at once apparent that this boat would not hold the six of us, so Dr. Vieta was sent back to Bayamo with the escort and the horses, while five of us prepared to run the gauntlet of Spanish guns and outwit Spanish gun-boats with a craft not much larger than a skiff and with sails of gunny-sacks!

There was a storm raging at the time we had fixed upon for our departure and we could not venture on the water while the waves were rolling so fiercely. Yet even in waiting there was danger! It was the time of the full moon and should the clouds dissipate with the passing of the gale our presence might be detected. But the fates were with us!

At 11 o'clock we embarked. With only five aboard the boat was well down in the water. The ragged clouds rushed like mad things across the face of the moon, alternately hiding and disclosing us, while four tugged at the oars and a fifth steered a course. We could not see the fort as we passed, and that perhaps was the reason we were not seen, but it required no great stretch of imagination to picture the frowning muzzles of the great guns and we toiled on, expecting at any moment to hear the boom of a cannon and the scream of a shot. Our little craft reeled and tossed like an egg-shell and many times we were on the point of capsizing, but our sailors knew the course, our gunnysack sails stood the test and soon we were making headway "across the trackless green".

Weary with the unwonted toil and with nothing to break the monotony of riding first one wave crest and then another, I fell asleep sitting bold upright. But not for long. An immense wave hit us, nearly filling our boat with water and almost capsizing us. From that time on there was no sleep for anyone. It was bail, bail, bail the long night through. Drenched with brine, weary and worn, we were glad enough to get a glimpse of the sun as it peered through the haze on the horizon.

"Un vapor, Senores!" (a steamer) cried the steersman.

A feeling of alarm agitated every heart. Suppose it should be a Spanish warship? That would mean short shrift for all of us.

"Dos vapores, tres vapores, Caramba! doce vapores!" cried the steersman, my companions echoing his cries. Could it be the Spanish fleet?

But no, it was the battleships of Admiral Sampson, steaming eastward to attack San Juan del Puerto Rico!

We breathed easier!

All that day we broiled and bailed, bailed and broiled. Yet no one slept or relaxed his anxious outlook. Despite the presence of the United States warships a gunboat might have escaped their vigilance and if so might overtake and capture us. Night fell on five of the most tired men that ever lived. We were almost worn out with fatigue, but for us there could be no rest. With the darkness came the wind again and with the wind the mighty waves and again it was bail, bail, bail, to keep the little vessel afloat. It was with feelings of intense relief that on the next morning, May 7th, at about 10 o'clock, we sighted the Curly Keys at the southern end of Andros Islands of the Bahama group and right gladly did we land there for a brief rest.

That afternoon we overhauled a sponging schooner, with a crew of thirteen negroes, who spoke some outlandish gibberish we did not understand, but sign language is universal, and soon we had made arrangements for a transfer. This schooner carried a litter of pigs for food and an accordeon. I never want to hear an accordeon again. Tired almost to the point of utter exhaustion, I vainly sought sleep but the shrill notes of that instrument prevented it.

Next afternoon we were captured by quarantine officials as we turned the east end of New Providence Island, and were incarcerated at Hog Island, the fiction of yellow fever in Cuba having given them the excuse.

But next day I got word to the American consul general, Mr. McLean, and on May 10th he arranged our release. May 11th the schooner Fearless drew near the wharf and we went aboard.

We had got in behind Florida Keys when luck deserted us. The wind went down and all day May 12th we lay becalmed, but at night a breeze came up and on the morning of May 13th we were in Key West.

That night we took a train for Tampa and there boarded a train for Washington. We arrived on schedule time and I reported to Russel A. Alger, secretary of war, who heard my story and told me to report to General Miles, taking General Garcia's aids with me. After he had received my report General Miles wrote to the secretary of war: "I also recommend that First Lieutenant Andrew S. Rowan, 19th U.S. Infantry, be made a lieutenant-colonel of one of the regiments of immunes. Lieutenant Rowan made a journey across Cuba, was with the insurgent army with Lieutenant-General Garcia, and brought most important and valuable information to the government. This was a most perilous undertaking, and in my judgment Lieutenant Rowan performed an act of heroism and cool daring that has rarely been excelled in the annals of warfare."

I attended a meeting of the cabinet a day or so after my return, in company with General Miles, and at the close I received President McKinley's congratulations and thanks for the manner in which I had communicated his wishes to General Garcia and for the value of the work.

"You have performed a very brave deed!" were his last words to me, and this was the first time it had occurred to me that I had done more than my simple duty, the duty of a soldier who " Is not to reason why", but to obey his orders.

I had carried my message to Garcia.

第四部分
上帝对你做了什么？

（马克·戈尔曼）

What's God Doing For You?

一百多年前的一天，一位编辑为了不让即将出版的杂志留下空白，于是写了一篇关于一名美国士兵的文章。这篇看起来并不重要的文章竟然奇迹般地成为出版史上销量最高的出版物之一，这就是《致加西亚的信》。它被翻译成多种文字，销量超过一亿册。究竟这篇文章蕴涵着怎样的价值？它在世界上引起如此轰动的原因是什么呢？

1899年，阿尔伯特·哈博德为小杂志《菲士利人》写了一篇文章。哈博德与他的家人在喝茶之时谈论关于美西战争的话题。人们都认为古巴起义军首领加西亚将军是个英雄，因为他在古巴战役中有着至关重要的作用。但是，哈博德的儿子博尔特却不这样认为。

“在我看来，”博尔特坦率地说，“战争中真正的英雄并不是加西亚将军，而是罗文中尉，就是给加西亚将军送信的那个人。”儿子的话让哈博德陷入了沉思。

为此，哈博德写了一篇文章，这就是《致加西亚的信》，这期杂志很快出版发行了。最初，他并没有关注这篇文章。直到请求杂志加印的要求逐日增多，他才不得不关注起来。

因为要求加印的请求太多了，致使杂志逐步陷入困境。面对接连不断的订单，哈博德非常迷惑，他想知道，人们为什么单单对这期杂志感兴趣呢。最后，他了解到，所有的订单都是朝着那篇关于罗文的文章而来的。这使他十分惊讶！面对十万份、五十万份，甚至高达一百万份的订单，哈博德疲于应付，只好将重印的版权卖给那些需要量极大的人，因为自己的印刷能力不足。为什么人们对这个默默无闻、名叫安德鲁·萨默斯·罗文的中尉如此感兴趣呢？原因就是：每个人都在寻找像罗文这样的人！

1895年，古巴人民正在为摆脱西班牙的统治而斗争。西班牙人占领了古巴岛，疯狂地剥削和奴役当地居民。所以，古巴人民想赶走西班牙人，重新获得自由。美国十分关注古巴的形势发展，因为古巴是美国的邻国，而且美国在古巴有经济投资。1897年，古巴境内的形势更加恶化，古巴民族主义者和西班牙士兵在哈瓦那大街发生了暴力冲突，并引发了大规模的暴乱。缅因号战舰被美国总统麦金莱调往古巴境内，停靠在哈瓦那港湾，成为美国政府在古巴境内的代表性标志。通过此事，美国政府明确地向西班牙政府传达了一个讯息，那就是美国一定会保护自己在古巴的利益。虽然缅因号战舰具有威慑西班牙人的作用，但它一直没有参加任何反西班牙的行动。

1898年2月15日，发生在哈瓦那的一次爆炸却使这艘战舰沉入海底。爆炸发生之地距离美国海岸不足100英里。美国人民将此次事件看作是西班牙人的公开挑衅。作为回击，麦金莱总统向西班牙政府下达了最后通牒：退出古巴。4月，美国和西班牙正式开战，这就是历史上的美西战争。这场战争不仅解放了古巴，还解放了菲律宾群岛。

与西班牙宣战前夕，麦金莱总统在接见美国军事情报局局长阿瑟·瓦格纳陆军上校时，问道："谁能帮我把信送给加西亚，在哪里可以找到这样的人？"打赢这场仗，美国与古巴起义军合作是关键。所以，迅速和起义军领袖克里克托·加西亚将军——一个地地道道的古巴克利奥尔人——取得联系就变得非常重要。那个时候，加西亚将军领导自己的队伍正在丛林中与敌人进行战斗。由于他是西班牙政府通缉的人物，因此他的行踪十分隐秘，谁也不知道他具体的位置。

面对总统的提问，瓦格纳上校不假思索地说："我有一个人选——安德鲁·萨默斯·罗文，一位年轻的中尉。假如有人能把信送给加西亚将军，那么一定是他。"

一个小时后，瓦格纳上校出现在罗文中尉的面前，对他说："年轻人，你必须将这封信送给加西亚将军，他可能隐藏在古巴东部的某个地方……你必须凭借自己的力量计划和安排一切行动。"然后，瓦格纳上校和罗文握手告别，再三叮嘱说："必须把信送给加西亚将军。"罗文接过信后没有问任何问题，就开始了寻找加西亚的行程。

最后，罗文成功地把信送到了加西亚将军手中，还为麦金莱总统带回了重要情报。当时，罗文接到信后并没有提出诸如此类的问题："加西亚在哪里？""他的相貌是什么样的？""如何与他取得联系？""怎样到达那里？"罗文只是服从命令，并做他应该做的事。

在我们中间有罗文这样的人吗？有不需要向上司提出任何疑问就成功地把信送给加西亚的人吗？有不需要上司指挥，就能独立完成工作的人吗？如果没有，恐怕上司得自己做所有的事了。

如果我让一个人独立去完成一项任务，等他再一次见到我时，他说的是："那项任务我已经完成了，还有需要我做的事吗？"这样的人在哪里？我怎样才能找到他？我可以找到一个罗文吗？有能把信送给加西亚的人吗?

这样的人存在，但数量不多。一些罗文们有可能就在阅读这篇文章，他们将会成为伟大的人物。伟大意味着超越了庸俗。这些人并不简单地做他人要求自己的事情，他们会超越他人的想象，追求完美。以下这些文字取自阿尔伯特·哈博德的《致加西亚的信》。虽然文章写于100多年前，但读起来仿佛刚刚写成的一样：

我要特别提出的是：麦金莱总统把一封写给加西亚的信交给罗文，罗文接过信后却没有问："他在哪里？"这种伟大的精神会永远流传世间。我们应该为拥有这种精神的人塑造永不腐朽的雕像，并将雕像放在每一所大学里。年轻人需要的不仅是书本知识，也不仅是认真听取各种教诲，

而是需要能让他们挺胸抬头的敬业精神。有了这种精神，他们会坚定自己的理想，迅速行动，集中精力，全力以赴地完成任务——“把信送给加西亚”。……如果你不认同我的结论，那就来做个测验。假设你有六名属下，在你办公时，你找来其中的一名，吩咐道：“马上在百科全书中查找科勒乔的资料，并做一份关于他生平的简短备忘录。”这名员工会不会迅速地回答你“是，先生”，然后马上去工作呢？实际上，他不会。他会用怀疑的眼光看着你，然后问你下面这些问题，或者是这些问题中的某些问题，或者是更多的问题。

科勒乔是什么人？

在哪一本百科全书中查找？

百科全书放在哪里了？

我的工作就是做这个吗？

查理为什么不能做这件事呢？

科勒乔去世了吗？

真的很紧急吗？

是否需要我把百科全书找来你自己查呢？

你想知道要查科勒乔的哪些资料呢？

……

如果你是个有智慧的人，你就不会不厌其烦地向你的属下解释提出的问题：应该在字母“C”的索引中查找科勒乔的资料，而不是在字母

"K"的索引中。你会平静地说"算了"然后亲自去查找。

一百多年过去了，人们并没有什么变化，不是吗？每一次我给他人布置任务时，总会听到无数的疑问。见此情景，我马上告诉自己："这个可怜的人不能把信送给加西亚。"

实际上的确如此，能把信送给加西亚的人并不多。更多的人选择了满足平庸的现状，对他们而言，达到平均水平就足够了。我并非不理解这种思维方式。只有你想要成功，你才可能获得成功；你成功，不是因为生活选择了你，而是因为你选择了生活。为自己作出选择，你可以选择得过且过的生活，也可以选择完美的生活。

《圣经·马可福音》中有这样一个故事：走过一段遥远的路程后，耶稣和他的弟子们十分疲劳，而且饥渴难耐。耶稣走到一棵无花果树前，却发现树上没有果实。失望之下，耶稣诅咒了这棵树。第二天，当他们再次路过这棵树时，一个弟子惊奇地发现，树已经枯死了。

最近，我再次读到这则故事，发现了一些我以前阅读的过程中没有发现的问题。文章里说，那棵无花果树之所以没有果实，是因为当时不是它结果的时节。所以，我的问题就是："上帝，难道你不认为自己对这棵树的惩罚太严厉了吗？因为在那个时节，任何一棵无花果树都不可能长出果实啊！"

就在当晚的凌晨两点，我忽然从床上坐了起来，因为上帝对我发话了："如果人们所做的一切都顺其自然，那人们就不会记得有我了。"

上帝的意思是，他不愿意看到我们只做那些自然而然的事情，以及简单和舒服的事情，他愿意看到我们超越这些。对于我们而言，顺其自然意味着平庸。平庸是上帝希望我们最后选择的一条路。通过无花果树的例子，耶稣告诉我们应该做什么，他希望那棵无花果树不仅仅是多结果实，而且一年四季都要结果。就像我们，为什么可以选择更好时总是选择平庸呢？如果你能在一年中的某一天结出果实，那一年有365天，你为什么不能完全利用呢？为什么我们要做别人正在做的事情？为什么我们不可以超越平庸？

如果一个人只做自然而然的事情，他不会在奥林匹克竞赛中取得胜利。那些获得金牌的运动员必须超越已有的成绩。我厌恶平庸，我的感觉和哈博德写下这段话时的感觉几乎一模一样：

最近，我发现很多人都对那些"收入微薄而且永无出头之日的人"和"为求得一份稳定工作四处奔波、无家可归的人"表示出深切的同情。同时，这些人还痛骂雇主。但是，从来没有人提到过，有些雇主尽自己最大的努力都没有能够使那些懒散的职员做些有意义的工作；也没有人提过，有些雇主是如何以自己的耐心去感动那些自己一转身就投机取巧的员工的。

……

我这样说是不是有些夸张呢？可能吧。不过，如果全世界都变成了贫民窟，那么，我要为成功者说几句同情的话——他们在几乎没有成功可能性的情况下全力引导其他人，并最终取得成功……我尊重那些不管老板是否在办公室都会勤奋工作的人。我也尊重那些把信交给加西亚的人，他们只会静静地接受命令，不会提出愚蠢的问题，不会出门后把信扔到水沟里，不会不去送信而做其他无关的事情。这些人永远不会被解雇，也永远用不着为了加薪而罢工。

文明，就是焦急地寻找这些人的一个长远过程。这些人不管要求什么东西，最后都会得到。他的才干如此独一无二，不可或缺，任何雇主都不愿失去他。他在每个城市、乡镇、村庄，以及每个办公室、商店、工厂都会受到热烈欢迎。世界需要这样的人才，需要能把信送给加西亚的人。

不要说他人对你的期望值远比你对自己的期望值高。如果他人发现了你工作中的不足之处，那你就必须承认，自己不是优秀的，也不用为此寻找理由和借口。承认你没有尽全力做到最好，千万不要积极地站出来为自己解释。当我们可以选择更好时，为什么要甘于平庸呢？我讨厌人们总是这样认为，以更高的标准要求自己并不是自己的天性。他们大多会说："我的性格与你不同，事实上，我没有你那么大的野心，那不是我的天性。"

对于这样的说法，我的回答就是："改变。"实际上，这只是一个决定问题，下决心去做一个改变的决定吧！

《圣经·马太福音》中有一篇文章阐述了这个主题，很有启示意义。文章的主要内容是这样的：一个人将要远行，在出发之前将自己的三个仆人召集起来，把自己的财产交给他们保管。他根据三个仆人的能力大小分配财产。第一个人仆人分到了5个塔兰特，第二个仆人分到了2个塔兰特，第三个仆人分到了1个塔兰特。第一个仆人拿着5个塔兰特去做生意，结果赚了5个塔兰特，他将主人的财产翻了一番；同样的，第二个仆人拿着2个塔兰特去做生意，结果赚了2个塔兰特，也将主人的财产翻了一番；第三个仆人将自己手中的那个塔兰特埋在了土里。

后来，主人回来了，开始和仆人们对账。第一个仆人一共拿出了10个塔兰特，得到了主人的夸赞："干得好，你是个优秀而又忠诚的人！以前分给你的任务太少了，我会让你管理更多的事务。现在我们一起来分享成功的荣耀吧！"

第二个仆人一共拿出了4个塔兰特，主人说："干得好，你是个优秀而又忠诚的人！以前分给你的任务太少了，我会让你管理更多的事务。现在我们一起来分享成功的荣耀吧！"

最后，第三个仆人两手空空地进来了，他说："主人，我了解您的性格，您意欲成为一个强大的人，收获没有播种的土地，收割没有撒种的庄稼。我很害怕，于是把钱埋在了地里。看那里，那儿埋着你的钱。"他的主人回答说："既懒惰又缺乏道德的仆人，你已经知道我想收获没有播种的土地，收割没有撒种的庄稼，那你为什么不把钱存到银行家手中？那样，我回来后不仅能拿到我的本钱，还可以得到本钱的利息。我还可以把这1个塔兰特，送给有10个塔兰特的人。那些已经拥有很多的人如果被给予，会更加富裕。对于那些穷困的人来说，他们所拥有的东西也会被剥夺。"

第三个仆人本以为自己会受到表扬，因为他妥善地保存了主人的财产。在他的观念里，没有将财产弄丢，也没有使财产减少，这就是圆满地完成了主人交给他的任务。但是，他的主人却不这样想。他不愿意看见仆人只做那些自然而然的事，他期盼能看到他们的创造性，他渴望他们能超越平庸。三个仆人中，只有两个人做到了。他们把主人留给自己保管的塔兰特的数量翻了一番，也就是说增加了百分之百！而那个最笨仆人的想法只是得过且过。

在我的一生中，我遇见过许多和第三个仆人持有相同态度的人，他们这样认为："只要完成那些不得不做的事情就可以了，我不想把每件事做得都很完美。"

你如何处理自己被赋予的任务？你只做你周围的人做的事情吗？你的态度是否和第三个仆人一样？

沃纳·冯·布劳恩是美国宇航局空间研究与开发中心的主设计师。此外，他还曾担任阿波罗4号专案的主设计师。他认为这项计划与土星5号火

箭（也称月球火箭）有密切关系，因为土星5号火箭的任务是推动太空船进入预定轨道。他说：“土星5号身上的部件总共有5 600 000个，就算它有99%的可靠性，那就意味着它身上可能有5 600个地方存在缺陷。不过，阿波罗4号计划在进行模拟飞行时，只发现了2处异常情况，这证明了其可靠性可达99.999%。假设一部汽车由13 000个部件组成，如果它具有和火箭同样的可靠性，那么，100年之后，汽车才会出现第一块有缺陷的部件。”

为什么土星5号火箭的可靠性要远远高于我们的汽车呢？因为美国宇航局追求完美，他们制定的标准比汽车工业的更严格。美国宇航局应该成为我们学习的榜样。上帝不是正希望我们这样做吗，为自己制定一个高于他人的标准。

你应该扪心自问：“我能把信交到加西亚的手中吗？如果我被告知他隐藏在古巴某处的丛林中，我能把信交到他手中吗？如果我不清楚他的相貌，不知道如何去找他，我能把信交到他手中吗？”如果你一心想要成功，那么，你一定会找到通往成功的道路。如果你排除万难，追求成功，那成功一定会到来。

现在，大多数人都是找借口的专家。对于不能下定决心去做的事，他们总有诸多的理由。为什么不能对自己的工作尽职尽责、尽善尽美呢？然而，人们经常对我说的，只是种种不能完成工作的借口。

成为罗文那样的人吧！作出了决定，就马上行动！当然，我们可能会被一些事情拖累；在前进的道路上，也可能会陷入困境。但是，为了完成自己的任务，我们必须坚持，直到到达胜利的终点。有些时候，我不知道自己还能不能超越他人，但无论如何，我都不会停下脚步，不会轻言放弃。逃避并不是你所能作出的唯一的选择。我一定要完成自己的任务，在生活的每一个领域，都做到十全十美。即使跌倒了，我也要再次爬起来。我会全力以赴，直至取得胜利！

上帝，将罗文这样的人赐给我们吧！

如果有人让我给加西亚送信，我想我肯定能送到。你也许会认为我这个人太自大了，但实际上，这并不是自大，而是自信。当你将一封信交给我，说"把信送给加西亚"，我相信自己能做到。我也想让你把信送给加西亚，而且要做到最好！假如有人说你一生都无所作为，这一定是谎言，你切不可相信。对你而言，他人所说的消极事件，与你没有任何关系。

下定决心，然后采取行动。成功是1%的灵感加上99%的汗水。只要你开始行动，并付出努力，你一定会成功。你下定决心出色完成工作了吗？把信送给加西亚，你准备好了吗？

在我办公室的墙上挂着一块匾，刻着如下的文字：

卓越就是比别人想得更多；冒更多的风险；有更多的梦想；有更高的期望。

选择一份完美的生活，努力达到目标，做自己想做的梦，你一定会成功！把信送给加西亚！

马克·戈尔曼

Over 100 years ago, a brief article was written to fill an empty space in a magazine which was otherwise ready for publication. This seemingly insignificant work, about a soldier in the U.S. Army, has since become one of the most published documents in the history of printed word. *A Message To Garcia* has been translated into every major language on earth, with over 100 million copies in print. What was the significance of this article, which caused such a stir around the world?

In 1899, a man by the name of Elbert Hubbard wrote an editorial for a small magazine called *The Philistine*. Over tea, Hubbard was discussing the Spanish-American War with his family. Everyone had been cheering General Calixto Garcia, the leader of the Cuban rebel forces, as the key to winning the war in Cuba, when Hubbard's son, Bert, put forth this argument.

"In my mind", ventured Bert, "the real hero of the war was not General Garcia, but Lieutenant Rowan, the man who got the message to Garcia." His son's words leaped in Hubbard's heart.

Hubbard wrote the article, *A Message To Garcia* and the edition went to print. He thought little more about it until the magazine began getting requests for reprints of that particular edition.

More and more requests for reprints came in until the magazine was literally swamped. Puzzled by the overwhelming number of orders, Hubbard asked why people were interested in that particular copy of the magazine. He was surprised to learn that the demand was for the "filler" article he had written about Rowan. Orders came in for 100 000 copies, 500 000 copies, 1 000 000 copies. Eventually, Hubbard was forced to simply grant permission to those who wanted large numbers of reprints, because of his limited ability to publish in those quantities. Why are so many people interested in an article about some unknown lieutenant by the name of Andrew Summers Rowan? The reason is: everyone is looking for individuals such as Rowan.

In 1895, the little island nation of Cuba was struggling to be free from Spanish rule. The Spanish soldiers who occupied the island oppressed and brutalized the people. They desperately wanted to be free. The United States had a strong interest in Cuba, not only because of its geographical proximity to the United States, but also because of our financial investments there. By 1897, the situation in Cuba had deteriorated to the point that there was rioting in the streets of Havana between nationalists and Spanish soldiers. President McKinley dispatched the battleship Maine as a visible indicator of the United States' presence in Cuba. The American battleship, sitting in Havana harbor, sent a clear signal to the Spanish government of our country's resolve to protect our interests in Cuba. Although a formidable presence, the Maine did not engage in any hostile act against Spain.

On February 15th, 1898, however, an explosion rocked the Havana harbor sinking the U.S. battleship. The American people were greatly alarmed over this open act of aggression less than 100 miles off our country's coast. McKinley sent an ultimatum to Spain to get out of Cuba. By April, the United States was at war with Spain. Ultimately, the Spanish-American War proved to liberate, not only the nation of Cuba, but the Philippine Islands, as well.

Just before declaring war, President McKinley was meeting with Colonel Arthur Wagner, head of the Bureau of Military Intelligence for the United States. "Where", asked President McKinley, " can I find a man who will carry a message to Garcia?" Cooperation between the rebel forces in Cuba and the United States was essential to the success of the campaign. It was vital to quickly communicate with the leader of the rebels, General Calixto Garcia, a Cuban-born Creole. General Garcia was somewhere in the mountains of Cuba leading the rebel troops in their fight for independence. He was a hunted man by the Spanish army. No one knew his exact whereabouts.

Colonel Wagner did not hesitate in his answer to the President. "I have a man—a young officer, Lieutenant Andrew Summers Rowan. If anybody can get a message to Garcia, Rowan can."

An hour later, Col. Wagner stood before Lieutenant Rowan. "Young man," said the superior officer, "you must carry a message to General Garcia, who will be found somewhere in the eastern part of Cuba. You must plan and act for yourself. The task is yours and yours only." Col. Wagner then shook Rowan's hand and repeated: "Get that message to Garcia." Without asking one question, Rowan left to find Garcia.

Rowan delivered the message to Garcia and the response got back to McKinley without Rowan ever asking, "Where is he? What does he look like? Who are his contacts? How do I get there?" He simply took the orders and did what he was asked to do.

Is there a Rowan among us? Is there somebody who can get a message to Garcia without having to do an interrogation of his senior officer first? Is there someone who can get the job done without needing to have his employer hold his hand until the task is completed? If not, the boss might as well do it himself.

Is there somebody that I can just ask to accomplish a task, and the next time I see them I am told, "I'm finished with that. What do you want me to do next?" Where can I find someone like that? Where is he? Can I find a Rowan? Is there someone who can get a message to Garcia?

They are out there. There's just not enough of them. There are probably some Rowans reading this right now. There will always be a few of those individuals who are extraordinary. Extraordinary means above ordinary. Those who don't just do what is expected of them; they surpass the expectations of others, in their pursuit of excellence. Here is an excerpt from Elbert Hubbard's article written over 100 years ago. It sounds as if it could have been written today:

The point I wish to make is this: McKinley gave Rowan a letter to be delivered to Garcia. Rowan took the letter and did not ask, "Where is he ?" By the eternal, there is a man whose form should be cast in deathless bronze and the statue placed in every college of the land. It is not book-learning young men need, nor instruction about this and that, but a stiffening of the vertebrae which will cause them to be loyal to a trust, to act promptly, concentrate their energies: do the thing—"Carry a message to Garcia!" You reader, put this matter to a test. You are sitting now in your office. Six clerks are within call. Summon any one and make this request: "Please look in the encyclopedia and make a brief memorandum for me concerning the life of Correggio." Will the clerk quietly

say "Yes, sir", and go do the task. On your life, he will not. He will look at you out of a fishy eye and ask one or more of the following questions:

Who was he?

Which encyclopedia?

Where is the encyclopedia?

Was I hired for that?

Don't you mean Bismarck?

What's the matter with Charlie doing it?

Is he dead?

Is there any hurry?

Shan't I bring the book and let you look it up yourself?

What do you want to know for?

...

Now if you are wise you will not bother to explain to your assistant that Correggio is indexed under the C's, not under the K's, but you will smile sweetly and say "Never mind", and go look it up yourself.

People haven't changed in the last 100 years, have they? Every time I give someone a task and they start asking me a hundred questions, I immediately say to myself: " This poor soul could not get a message to Garcia."

Those who can get a message to Garcia are rare. The majority is satisfied with the status quo-with simply being average. I don't understand that mentality. I can't comprehend the paradigm of being satisfied with average. You are going to succeed because you decide to succeed. You are going to succeed because you make the choice that you will not let life choose for you. I will choose for myself. You can choose to live a life of "barely making it through" or choose a life of excellence.

I am reminded of an incident found in the *Bible* in the book of *Mark*. Jesus and his disciples had been traveling and were hungry. Jesus walked over to a beautiful fig tree, but it bore no figs. Jesus cursed the tree for not producing fruit. On the next day, as they passed by, one of the disciples noted that the fig tree had already withered and died.

Recently, while reading this story, I noted something, which I had overlooked on all of my previous readings. The scripture says that the tree was barren because figs were not in season. My obvious question was:"Lord, weren't you a bit harsh in your judgment of the fig tree? No trees had figs at that time of the year."

Later that same night, at 2 a.m., I sat straight up in bed as God spoke to me. He said:"If all you do is what comes naturally, I'm not impressed."

God does not expect us to do only what comes naturally. He expects us to do far beyond that which is convenient and comfortable. For us to remain in the natural flow of our existence is mediocrity. Average is the last thing God wants you and me to be. Jesus used the fig tree as an example of what he wants of us. He expected the tree to be productive and to bear fruit year round. Why settle for mediocrity when you have the option to be better than most? If you can produce one day out of the year, why not produce 365 days a year? Why do we have to do only what everyone else is doing? Why can't we be above average?

Nobody ever won an Olympic event by doing what came naturally. The athlete who will take home the gold must push beyond the limits of what has already been done. I am tired of average. I feel as Hubbard felt when he penned these words:

We have recently been hearing much maudlin sympathy expressed for the "downtrodden denizen of the sweatshop" and the "homeless wanderer searching for honest employment", and with it all often go many hard words for the men in power. Nothing is said about the employer who grows old before his time in a vain attempt to get frowsy never-do-wells to do intelligent work; and his long patient striving with "help" that does nothing but loaf when his back is turned.

...Have I put the matter too strongly? Possibly I have, but when all the world has gone a-slumming, I wish to speak a word of sympathy for the man who succeeds—the man who, against great odds, has directed the efforts of others, and having succeeded...My heart goes out to the man who does his work when the boss is away as well as when he is at home. And the man, who, when given a letter for Garcia, quietly takes the missive, without asking any idiotic questions, and with no lurking intention of chucking it into the nearest sewer, or of doing anything else but deliver it, never gets laid off, nor has to go on a strike for higher wages.

Civilization is one long anxious search for just such individuals. Anything such a man asks shall be granted. His kind is so rare that no employer can afford to let him go. He is wanted in every city, town, and village; in every office, shop, store and factory. The world cries out for such: He is needed and needed badly, the man who can "carry a message to Garcia".

Don't ever let it be said that someone else expected more of you than you expected of yourself. If anyone finds fault in a job which you have done that is less than excellent, don't make excuses. Admit that it was not your best. Don't stand up and try to defend yourself. Why settle for average, when excellence is an option? I'm weary of people saying that it's not in their nature to demand more of themselves. They may say: "My personality is different than yours. I'm not as aggressive as you are. It's not my nature."

My answer to them is "Change". Really, it's just a decision away. Make a decision to change.

A profound study on the subject of excellence can be found in the *Bible*. The book of *Matthew* tells of a man preparing for a trip to a far country who gathered his servants and entrusted his goods to them. The scripture says that to one, he gave five talents; to another he gave two; and to another, one. To each, he gave according to their own ability. He who had received the five talents went and traded them and made another five talents. Likewise, he who had two went and gained two more also. But, he who had received one talent went and buried his lord's money.

After a long time, the lord of those servants came and settled accounts with them. He who had received five talents came and brought another five talents. His lord said: "Well done, thou good and faithful servant. You have been faithful over a few things. I will make you ruler over many things. Enter now into the joy of your lord."

He also, who had received two talents, came and brought two talents more. His lord said: "Well done, thou good and faithful servant. You have been faithful over a few things. I will make you ruler over many. Enter into the joy of your lord."

Then he who had received one talent came and said: "Lord, I knew you to be a hard man, reaping where you have not sown, and gathering where you have not scattered seed. I was afraid and went and hid your talent in the ground. Look there, have what is yours." His lord answered and said unto him: "Wicked and lazy servant, you knew that I reap where I have not sown and gather where I have not scattered seed. So, you ought to have deposited money with the bankers and at my coming I would have received back my own with interest. Therefore, take the talent from him and give it to him that has ten talents. To everyone who has, more will be given and he will have abundance; but from him who does not have, even what he has will be taken away."

This servant thought he would gain his master's approval by not losing the one talent that had been given him. He thought he had accomplished something by not having lost it or gambled it away. The master, however, saw it differently. The master was not expecting his servants to do only what came naturally. He wanted them to excel. He wanted them to go beyond what was average. Two of them did. They doubled what he gave to them—a 100% increase. The foolish servant's mentality was to "just get by".

I have met so many people in my lifetime who have this attitude: "Let me do only what I absolutely have to do to make it through. I am not going to do anything with excellence."

What are you doing with what you have been given? Are you only producing as much as everyone else around you? Is your mindset like that of the foolish servant?

Werner Von Braun, head engineer of NASA's Space Research and Development for the Apollo IV Project said this concerning the Saturn V rocket, which was used to propel the spacecraft for that mission. "The Saturn V has 5 600 000 parts. Even if we had a 99% reliability, there would still be 5 600 defective parts. Yet, the Apollo IV mission flew a textbook flight with only two anomalies occurring, demonstrating a reliability of 99.999%. If an average automobile with 13 000 parts were to have the same reliability, it would have its first defective part in about a hundred years."

Why aren't our automobiles built with the same precision as the Saturn V rocket? Because NASA holds themselves to a higher set of standards than the automobile industry. We need to be like NASA. God is looking for us to pursue excellence—to set a higher standard for ourselves than everyone else sets.

I want you to ask yourself: "Could I get a message to Garcia? If I were told that he was hidden somewhere in the jungles of Cuba, could I get a message to him? If I didn't know what he looked like, or where to find him, could I do it?" If you are desperate to succeed, you will find a way. If you purpose in your heart to succeed, you will!

We have become experts with excuses—of why we can't do what we are supposed to do. Why can't we just take a job and do it with excellence? People tell me all kinds of excuses of why they can't do what they're supposed to do.

Be a Rowan. Do it! Just make a decision. Make a choice. Something may slow me down. I may get bogged down in my tracks. There may be times I find myself drowning in quicksand, times I have to hang on to make it through, times when I feel so downtrodden, I don't know if I can put one foot in front of the other, but I will not quit. I will not give up. Quitting is not even an option. I will accomplish the task that is set before me. I will pursue excellence in every area of my life. Even though I may fall down, I will get back up. I will dust myself off, and keep pressing on until I win.

God, give us people like Rowan!

If I were asked to get a letter to Garcia, I know I could. You may think that's arrogance on my part, but it's not. It's confidence. I know that if you handed me a letter and said "Get this to Garcia", I could get it there. I want you to get a message to Garcia, too. Be the best! If you have been told all your life that you cannot achieve, don't listen to those lies. It doesn't matter what negative things others may have told you.

Make a decision. Success is 1% inspiration and 99% perspiration. If you will just put in the effort, you can do it. Are you willing to make the decision to get the job done with excellence? Are you prepared to carry the message to Garcia?

Hanging on my office wall is a plaque with this inscription:

Excellence is the result of caring more than others think is wise; risking more than others think is safe; dreaming more than others think is practical and expecting more than others think is possible.

Choose to live a life of excellence. Pursue the goal. Dream the dream. You can do it. Get the message to Garcia!

—by Mark Gorman

第五部分

一本可怕的书

（威廉·亚德利）

It Said Everything

对于一个管理者来说，《致加西亚的信》能够给自己的团队传递一些重要的启示。

塔拉哈西（佛州首府）——在杰布·布什当选州长（美国佛罗里达州州长，第43任总统（2001—2009）乔治·沃克·布什的弟弟）的那天，他在一本毫不起眼的小书里签上了自己的名字，并把书送给了自己新的副手。

这本只有支票簿大小的书就是《致加西亚的信》，现在还摆放在弗兰克·布洛根的副州长（任期为1999—2003年，现已卸任）办公室的一张茶几上。布什在签名时，写下了这样一句话："你是一个信使！"

后来，布洛根成为杰布·布什政府里行政部门的信使之一。

有一段时间，政府新闻机构里的工作人员在墙上贴了一张纸，谁读过《致加西亚的信》这本书，谁就在纸上签下自己的名字。时间不长，这张纸上已经满满地全都是人们的签名了。

布什曾在回复一封电子邮件时说：“我把这本书献给那些在政府成立之初就和我们一起努力的人，我寻找那些能把信送给加西亚的人，让他们加入我们的队伍中。这些人无须监督就能努力工作，他们有着忠诚和坚毅的性格，是真正能够改变世界的人！”

实际上，《致加西亚的信》字数并不多，而且包装简陋。

《致加西亚的信》第一次出版的时间是1899年。文中讲述了忠诚而又勇敢的中尉安德鲁·萨默斯·罗文在1898年进入古巴寻找加西亚的事迹。在美国向西班牙宣战前夕，为了得到更多的情报，麦金莱总统派遣罗文去寻找加西亚将军。加西亚是古巴反对西班牙统治的起义军首领，美国只有与他联合才能保证对西班牙战争的胜利。罗文接过信后没有提出任何疑问就上路了。他最终把信送给了加西亚，并顺利回到华盛顿，带回了关于起义军以及敌人的情报。这些情报对于美西战争前夕美国的战争决策有不可估量的作用。

报纸对罗文的事迹进行了报道，并高度评价了他的行为。于是，罗文一时间名声大振。

纽约北部有个印刷商，也就是阿尔伯特·哈博德，他在这件事中获得了灵感，于是写了一篇关于罗文的文章。谁也没有想到的是，当初一篇不起眼的文章竟然在日后成为世界上销量最好的出版物之一。这本书成为无数雇主激励自己员工的教材。一个世纪过去了，这本书又成为杰布·布什和他年轻工作人员的人生信条。

在文中，阿尔伯特·哈博德写下了这样的话：

我要特别提出的是：麦金莱总统把一封写给加西亚的信交给罗文，罗文接过信后却没有问“他在哪里？”这种伟大的精神会永远流传世间。我们应该为拥有这种精神的人塑造永不腐朽的雕像，并将雕像放在每一所大学里。年轻人需要的不仅是书本知识，也不仅是认真听取各种教诲，而是需要那些能让他们挺胸抬头的敬业精神。有了这种精神，他们会坚定自己的理想，迅速行动，集中精力，全力以赴地完成任务——“把信送给加西亚”。

这本书不是一本简单的歌颂英雄的赞歌，而是一本成功的励志书。布什的新闻主管贾斯汀·赛非说过这样的话：“新闻机构里的每一个人都要仔细阅读这篇文章。文章为我们提供了一个绝妙的指导原则。我就经常用这个原则指导自己的行动：在完成任务时，不要被困难挡住前进的脚步；完成任务时不依赖他人。我认为，高级官员都应该熟悉这篇文章。”

现在，告诉你杰布·布什是如何读到《致加西亚的信》这本书的。

肯·瑞特是奥兰多的一名律师，曾为布什家族及老布什总统服务过。1998年，他将这本书推荐给杰布·布什。当时，布什正在竞选佛罗里达州州长。

瑞特曾写下这样的话："我从来都不让自己去埋怨什么。我人生的信条就是：为自己拥有的工作而全力以赴。"瑞特对自己推荐这本书给布什时的情景记忆犹新。

当我把这本书送给杰布时，杰布说："我对新时代的东西毫无兴趣。"

我不得不解释说："杰布，这本书文字不多，用喝一杯咖啡的时间阅读就足够了。而且，它不是新时代的东西，它几乎与我们历史一样悠久。"

当我再一次看见他时，他已经将这本书读完了。与我想象的一样，他告诉我："这是一本可怕的书，它讲明了一切！"

——威廉·亚德利

For the governor, *A Message to Garcia* imparts an important message to his troops.

TALLAHASSEE—The day he became governor, Jeb Bush signed the

inside of a small hardback book and presented it to his new No. 2 man.

Slim and barely bigger than a checkbook, the copy of *A Message to Garcia* now sits on an end table in the office of Lt. Gov. Frank Brogan. Bush wrote four words above his signature: "You are a messenger!"

Brogan, it turns out, is one of many messengers in the Bush administration.

For months, staffers in the governor's press office signed a sheet of paper tacked to a wall asking for the names of everyone who had read *A Message to Garcia*. By spring of this year the sheet was full.

"I gave it to all of the folks that started out with us at the beginning of the administration," Bush said recently in response to an e-mail." I look for people who will take the message to Garcia to be part of our team. People of determination and integrity that don't need too much adult supervision are the ones that can change the world!"

A Message to Garcia is an essay, really, 24 paragraphs simply bound and covered!

Originally published in 1899, *A Message to Garcia* recounts the intrepid loyal Lt. Andrew Summers Rowan undertook into the hills of Cuba in 1898. The United States would soon be at war with Spain, and President William McKinley sent Rowan to find Gen. Calixto Garcia, leader of insurgent Cuban forces fighting Spanish control. Without asking how or where he might find Garcia, Rowan set off. He found him, and returned to Washington to inform

McKinley of the strength and position of the rebels and the Spaniards. The information was crucial on the eve of war.

Newspapers celebrated the unlikely mission. Rowan became famous.

Inspired by the story, a printer in upstate New York wrote an essay that, in its day, became one of the best-selling publications in the world: motivational material for a generation of employers and, a century later, something of a creed for Bush and his young Republican staff.

The printer and essayist, Elbert Hubbard, wrote:

"The point I wish to make is this: McKinley gave Rowan a letter to be delivered to Garcia; Rowan took the letter and did not ask 'Where is he at?' By the Eternal! There is a man whose form should be cast in deathless bronze and the statue placed in every college of the land.

"It is not book-learning young men need, nor instruction about this and that, but a stiffening of the vertebrae which will cause them to be loyal to a trust, to act promptly, concentrate their energies: do the thing—'Carry a message to Garcia!'"

They do not chant the phrase like a fight song, but, Bush's Communication Director Justin Sayfie said, "Everyone in the press office was required to read the essay. It's a good guiding principle that I always use: Don't get bogged down in the obstacles to your task. Accomplish the task and do so using self-reliance. All senior staff have read it."

And how did Jeb Bush come to read A *Message to Garcia*?

Ken Wright, an Orlando attorney who has worked on campaigns for Bush and his former-president father, gave him a copy during his 1998 campaign for governor.

Wright interprets the essay this way: "Whining isn't allowed. That's my ethic: You got a job to do, do the job." He remembers exactly the conversation he had with the candidate when he recommended the book:

"I gave it to Jeb and Jeb said, 'I'm really not into this new age stuff.'

"I said: 'Jeb, read the book. It will take you a cup of coffee to read this book. This is not new age stuff. This is older than the hills.'

"When I ran into him next he had read it. His reaction was the same as I would have expected it be, 'That book was awesome. It says it all.'"

—by William Yardley

第六部分 人物简介

Character Introduction

阿尔伯特·哈博德

（Elbert Hubbard，1856—1915）

阿尔伯特·哈博德是19世纪末20世纪初杰出的哲学家、作家、演说家，也是一名成功的商人。1895年，他在纽约东奥罗拉创办了一家出版社，起名为罗伊克罗斯特。在当时，这也是一个艺术家和手工艺者的半公开团体，生产和销售种类繁多的手工艺制品。后来，他还成立了印刷厂和装订厂。哈博德创办了《菲士利人》月刊，在上面阐述自己的观点和主张。他的《短暂的旅行》也备受人们的欢迎。1899年，他以美西战争中罗文的事迹为题材，写了一篇关于效率和决心的文章，即《致加西亚的信》，引起了人们的广泛关注和阅读。1915年5月7日，他和妻子乘坐“露西塔尼亚”号船去英格兰旅行，因为船被击沉，夫妻俩不幸遇难。

Elbert Hubbard

加西亚

（Garcia，1836—1898）

克里克托·加西亚是古巴著名的革命家，也是带领古巴人民反对西班牙统治、争取民族自由的伟大领袖。他的活动招致了西班牙当局的痛恨，1878年底，他被逮捕并关进监狱。他在监狱里几进几出，但从来没有放弃过为自由而奋斗的信念。作为古巴起义军的首领，他在美西战争中发挥了至关重要的作用。1898年，他曾经到过美国，与麦金莱总统商讨战事。但不久，加西亚即因肺炎在华盛顿去世。

安德鲁·萨默斯·罗文
（Andrew S. Rowan，1857—1943）

罗文是美国军队里的一名军官。他于1881年从西点军校毕业。在美西战争中，罗文因将麦金莱总统的信送到加西亚将军手中而广受赞誉。战争结束后，他在菲律宾服役，后来又回到美国本土服役。1909年，罗文从军队中退役。1943年，罗文去世。

Elbert Hubbard (1856—1915)

Elbert Hubbard was a renowned philosopher, author, editor and lecturer of the late nineteenth and early twentieth centuries. In 1895, he founded the Roycrofters, a semi-communal community of artists and craftspeople, in East Aurora, NY. Here various handicraft products were made and sold, and here he established a printing press and bookbindery. His small magazine, the *Philistine*, carried his ideas to thousands, and his *Little Journeys* to The Homes of Various Famous Men were very popular. His *message to garcia* (1899), an inspirational piece on efficiency and determination based on a feat of A. S. Rowan, was widely read and quoted. He and his wife were lost at sea, May 7th, 1915, while traveling to England aboard the ill-fated Lusitania.

Garcia(1836—1898)

Calixto Inigues Garcia was a Cuban revo- lutionist and a leader in the Cuban insurrection against Spain. He was captured and imprisoned for his activities until its end in 1878. After his release he was again arrested. In 1895, he came to the United States and as the leader of the Cuban Insurgents, played an important role in the United States war with Spain. He died of pneumonia in Washington, D.C. in 1898 while there as part of a committee to discuss Cuban affairs with President McKinley.

Andrew Summers Rowan (1857—1943)

Rowan was an American Army officer and graduate of West Point class of 1881. After his service in the Spanish American War, he served in the Philippines and posts in the United States, retiring in 1909. He died in 1943.

第七部分

阿尔伯特·哈博德的商业信条

Elbert Hubbard's business "Credo"

我相信自己。

我相信自己销售的商品。

我相信自己就职的公司。

我相信自己的同事和助手。

我相信美国的商业方式。

我相信生产者、发明者、制造者、发行者，以及为自己所拥有的工作

而努力奋斗的人。

我相信真理的价值。

我相信人应该拥有愉悦的心情和强壮的体魄。我意识到，实际上，成功最重要的不是赚取金钱，而是创造价值。报酬迟早会来，只是需要一个过程而已。

我相信阳光、新鲜的空气、菠菜、苹果酱、微笑、酪乳、婴儿、丝绸和雪纺。时刻谨记，“满足”是英语中最伟大的词汇。

我相信我售出一件产品，就会交到一个朋友。

我相信当我和一个人分别时，我会做到：当我们重逢的时候，他非常高兴再次与我相见，我也非常高兴与他再次相见。我相信正在工作的双手，正在思考的大脑，以及充满爱的心灵。

阿门！

I believe in myself.

I believe in the goods I sell.

I believe in the firm for whom I work.

I believe in my colleagues and helpers.

I believe in American business methods.

I believe in producers, creators, manufacturers, distributors, and in all industrial workers of the world who have a job, and hold it down.

I believe that Truth is an asset.

I believe in good cheer and in good health, and I recognize the fact that the first requisite in success is not to achieve the dollar, but to confer a benefit, and that the reward will come automatically, and usually as a matter of course.

I believe in sunshine, fresh air, spinach, applesauce, laughter, buttermilk, babies, bombasine and chiffon, always remembering that the greatest word in the English language is "Sufficiency".

I believe that when I make a sale I make a friend.

And I believe that when I part with a man I must do it in such a way that when he sees me again he will be glad—and so will I. I believe in the hands that work, in the brains that think, and in the hearts that love.

Amen, and Amen!

书目

001. 唐诗
002. 宋词
003. 元曲
004. 三字经
005. 百家姓
006. 千字文
007. 弟子规
008. 增广贤文
009. 千家诗
010. 菜根谭
011. 孙子兵法
012. 三十六计
013. 老子
014. 庄子
015. 孟子
016. 论语
017. 五经
018. 四书
019. 诗经
020. 诸子百家哲理寓言
021. 山海经
022. 战国策
023. 三国志
024. 史记
025. 资治通鉴
026. 快读二十四史
027. 文心雕龙
028. 说文解字
029. 古文观止
030. 梦溪笔谈
031. 天工开物
032. 四库全书
033. 孝经
034. 素书
035. 冰鉴
036. 人类未解之谜（世界卷）
037. 人类未解之谜（中国卷）
038. 人类神秘现象（世界卷）
039. 人类神秘现象（中国卷）
040. 世界上下五千年
041. 中华上下五千年·夏商周
042. 中华上下五千年·春秋战国
043. 中华上下五千年·秦汉
044. 中华上下五千年·三国两晋
045. 中华上下五千年·隋唐
046. 中华上下五千年·宋元
047. 中华上下五千年·明清
048. 楚辞经典
049. 汉赋经典
050. 唐宋八大家散文
051. 世说新语
052. 徐霞客游记
053. 牡丹亭
054. 西厢记
055. 聊斋
056. 最美的散文（世界卷）
057. 最美的散文（中国卷）
058. 朱自清散文
059. 最美的词
060. 最美的诗
061. 柳永·李清照词
062. 苏东坡·辛弃疾词
063. 人间词话
064. 李白·杜甫诗
065. 红楼梦诗词
066. 徐志摩的诗

067. 朝花夕拾
068. 呐喊
069. 彷徨
070. 野草集
071. 园丁集
072. 飞鸟集
073. 新月集
074. 罗马神话
075. 希腊神话
076. 失落的文明
077. 罗马文明
078. 希腊文明
079. 古埃及文明
080. 玛雅文明
081. 印度文明
082. 拜占庭文明
083. 巴比伦文明
084. 瓦尔登湖
085. 蒙田美文
086. 培根论说文集
087. 沉思录
088. 宽容
089. 人类的故事
090. 姓氏
091. 汉字
092. 茶道
093. 成语故事
094. 中华句典
095. 奇趣楹联
096. 中华书法
097. 中国建筑
098. 中国绘画
099. 中国文明考古
100. 中国国家地理
101. 中国文化与自然遗产
102. 世界文化与自然遗产
103. 西洋建筑
104. 西洋绘画
105. 世界文化常识
106. 中国文化常识
107. 中国历史年表
108. 老子的智慧
109. 三十六计的智慧
110. 孙子兵法的智慧
111. 优雅——格调
112. 致加西亚的信
113. 假如给我三天光明
114. 智慧书
115. 少年中国说
116. 长生殿
117. 格言联璧
118. 笠翁对韵
119. 列子
120. 墨子
121. 荀子
122. 包公案
123. 韩非子
124. 鬼谷子
125. 淮南子
126. 孔子家语
127. 老残游记
128. 彭公案
129. 笑林广记
130. 朱子家训
131. 诸葛亮兵法
132. 幼学琼林

133. 太平广记
134. 声律启蒙
135. 小窗幽记
136. 孽海花
137. 警世通言
138. 醒世恒言
139. 喻世明言
140. 初刻拍案惊奇
141. 二刻拍案惊奇
142. 容斋随笔
143. 桃花扇
144. 忠经
145. 围炉夜话
146. 贞观政要
147. 龙文鞭影
148. 颜氏家训
149. 六韬
150. 三略
151. 励志枕边书
152. 心态决定命运
153. 一分钟口才训练
154. 低调做人的艺术
155. 锻造你的核心竞争力：保证完成任务
156. 礼仪资本
157. 每天进步一点点
158. 让你与众不同的 8 种职场素质
159. 思路决定出路
160. 优雅——妆容
161. 细节决定成败
162. 跟卡耐基学当众讲话
163. 跟卡耐基学人际交往
164. 跟卡耐基学商务礼仪
165. 情商决定命运
166. 受益一生的职场寓言
167. 我能：最大化自己的 8 种方法
168. 性格决定命运
169. 一分钟习惯培养
170. 影响一生的财商
171. 在逆境中成功的 14 种思路
172. 责任胜于能力
173. 最伟大的励志经典
174. 卡耐基人性的优点
175. 卡耐基人性的弱点
176. 财富的密码
177. 青年女性要懂的人生道理
178. 倍受欢迎的说话方式
179. 开发大脑的经典思维游戏
180. 千万别和孩子这样说——好父母绝不对孩子说的 40 句话
181. 和孩子这样说话很有效——好父母常对孩子说的 36 句话
182. 心灵甘泉